PARTICLE PHYSICS

BRICK BY BRICK

To Mum and Dad, thanks for all of your encouragement as a kid!

Published by Firefly Books Ltd. 2018

Text copyright © 2017 Ben Still
Design and layout copyright © 2017 Octopus Publishing Group Ltd.

All rights reserved. No part of this publication may be reproduced, stored in a retrieval system, or transmitted in any form or by any means, electronic, mechanical, photocopying, recording or otherwise, without the prior written permission of the Publisher.

First printing

Publisher Cataloging-in-Publication Data (U.S.)

Library of Congress Cataloging-in-Publication Data is available

Library and Archives Canada Cataloguing in Publication

Still, Ben, Dr., author
Particle physics brick by brick : atomic and subatomic physics explained...in Lego / Dr. Ben Still.
Includes index.
ISBN 978-0-228-10012-6 (softcover)
1. Particles (Nuclear physics)--Popular works. I. Title.
QC793.26.S75 2017 539.7'2 C2017-905112-1

Published in the United States by
Firefly Books (U.S.) Inc.
P.O. Box 1338, Ellicott Station
Buffalo, New York 14205

Published in Canada by
Firefly Books Ltd.
50 Staples Avenue, Unit 1
Richmond Hill, Ontario L4B 0A7

Printed and bound in China

First published by Cassell,
a division of Octopus Publishing Group Ltd
Carmelite House
50 Victoria Embankment
London EC4Y 0DZ

LEGO® is a trademark of the LEGO Group, which does not sponsor, authorise or endorse this book.

Ben Still asserts the moral right to be identified as the author of this work.

Editorial Director Trevor Davies
Senior Designer Jaz Bahra
Designer Paul Shubrook
Photographer Richard Clatworthy
Picture Researcher Giulia Hetherington
Senior Production Manager Peter Hunt
Managing Editor Sybella Stephens
Project Editor Sarah Green

PARTICLE PHYSICS BRICK BY BRICK

ATOMIC AND SUBATOMIC PHYSICS EXPLAINED... IN LEGO®

DR. BEN STILL

FIREFLY BOOKS

CONTENTS

SCIENTIFIC MODELS

When asked 'what is science?' I draw a deep breath. There are many answers that could be given; a historical one, a philosophical one, a practical one. Instead my answer is that science, at its core, is the search and desire for the most accurate possible analogy of Nature. Scientists build these analogies not in words but in the language of mathematics. The analogies they build aim to become the truest mathematical models of the way the universe works and where everything in it came from. But, like the words of a poet, the language we use cannot perfectly capture the true beauty of Nature.

Poets and scientists differ, however, when evolving their work. Re-writing of a poem is likely to be subject to differing human opinion, re-writing scientific models of Nature is not. Science answers only to hard and repeatable evidence from experiment. If time and again a new scientific model does not hold up in the face of new experimental results, then it is discarded. To progress, a new or modified model of Nature is required. In this way science has evolved, developing ever-more accurate mathematical models of Nature.

Experimental data is constantly reminding scientists of the shortcomings of their current mathematical models. Although our scientific understanding of Nature is imperfect, we embrace and quantify this imperfection in the error envelope of the measurements. What has been shown historically is that new science usually lurks in the detailed understanding of these errors.

When explaining complex ideas, we often find ourselves using our own analogies and models. This is especially true when communicating science which is abstract to our everyday experience of the world. Plastic bricks are of course not particles. Each brick is made from trillions upon trillions of particles. Yet I feel that they can be used to provide a fun and engaging analogy to our understanding of that subatomic world. This plastic brick analogy, by the very definition of the word, is not a perfect description of Nature, but does take us close to a complete picture of the Universe at the smallest scale.

THE STANDARD MODEL OF PARTICLE PHYSICS

The Standard Model is our current best model of the world at the smallest scales and was developed primarily throughout the 1960s and 70s. To this day there have only been minor adjustments made to it and every experiment which has put it to the test has shown it to be correct. Yet we know it cannot be a complete model of Nature. The Standard Model has no tested analogy for the dark matter which defines the size, shape and distribution of galaxies (see page 164). It also fails to explain the strange dark energy which is accelerating the expansion of our Universe. Most frustratingly for particle physicists is that the Standard Model is not able to tell us where all particles came from at the beginning of time itself (see chapter 8).

The Standard Model describes the properties and interactions between a collection of particles which are, as far as we understand, fundamental. Each one cannot be subdivided into anything smaller, so they are the truest building blocks from which our Universe is made. It does not tell us what the particles are made from; it does not tell us their size; and it cannot predict the strength of the forces which shape our Universe. It is a mathematical model which is designed to fit the experimental data seen – and it does this fantastically well.

The possible lives of Standard Model particles are explained not as solid particles but as extended objects called fields which stretch out to infinity in space and time. At each point in space and time a particle has a non-zero probability of interacting with other particles. Yet when the particles interact, all of the possibilities crystallize into a single point in space and time with set behaviour. It is this single point that best fits the idea of a particle as a solid little ball. Without discussing the maths in detail it is difficult to explain the field-like nature of particles. Here instead we use bricks to represent these point-like particles when they interact, although the field-like aspects will lead us to some seemingly illogical outcomes.

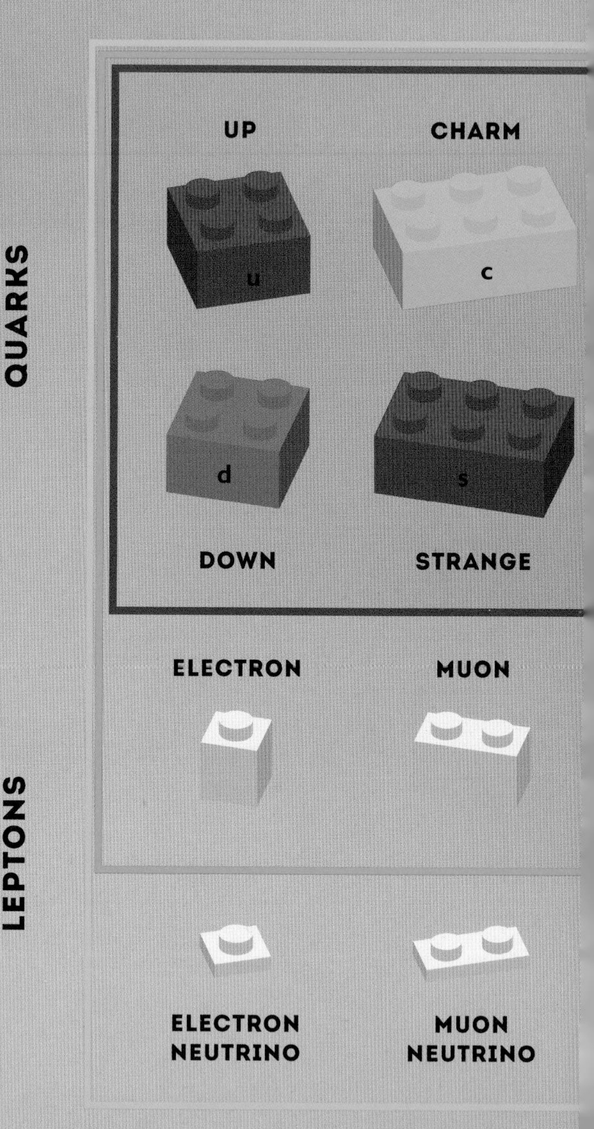

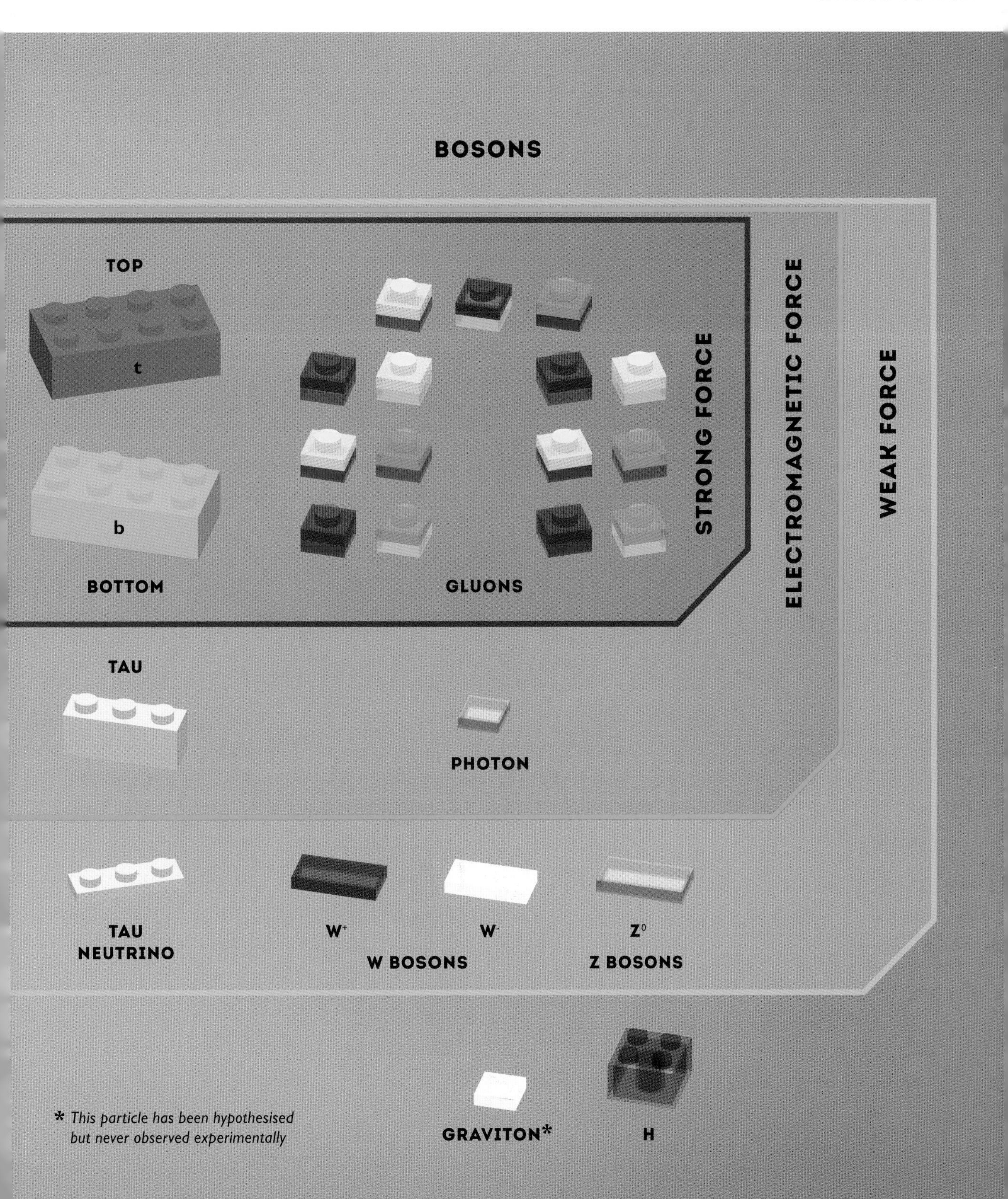
BOSONS
TOP
t
b
BOTTOM
GLUONS
STRONG FORCE
ELECTROMAGNETIC FORCE
WEAK FORCE
TAU
PHOTON
TAU NEUTRINO
W+
W-
W BOSONS
Z0
Z BOSONS
* This particle has been hypothesised but never observed experimentally
GRAVITON*
H

THE HISTORY OF PARTICLE PHYSICS

ELECTROMAGNETISM

QUANTUM PHYSICS

PREDICTIONS

1803
JOHN DALTON
Matter is made from different weight atoms
PAGE 14

1905
ALBERT EINSTEIN
Light is made of photons

1928
PAUL DIRAC
Dirac equations predict the antielectron and all other antiparticles
PAGE 88

1930
WOLFGANG PAULI
Neutrinos are needed to make sense of beta decay
PAGE 123

1935
HIDEKI YUKAWA
Mesons exchange the strong force in atomic nuclei
PAGE 92

DISCOVERIES

1897
J J THOMSON
Discovery of the electron, the first subatomic particle

1899
ERNEST RUTHERFORD
Discovery of alpha and beta radiation
PAGE 86

1911
EDWARD MARSDEN, HANS GEIGER AND ERNEST RUTHERFORD
Discovery of the atomic nucleus
PAGE 86

1919
ERNEST RUTHERFORD
Discovery of the proton

1932
JAMES CHADWICK
Discovery of the neutron

1935
CARL D ANDERSON
Discovery of the muon in cosmic rays
PAGE 89

1947
CECIL POWEL
CÉSAR LATTE
AND GIUSEPP
OCCHIALINI
Discovery of the pior
Yukawa's meson
PAGE 92

RADIOACTIVITY

COSMIC RAYS

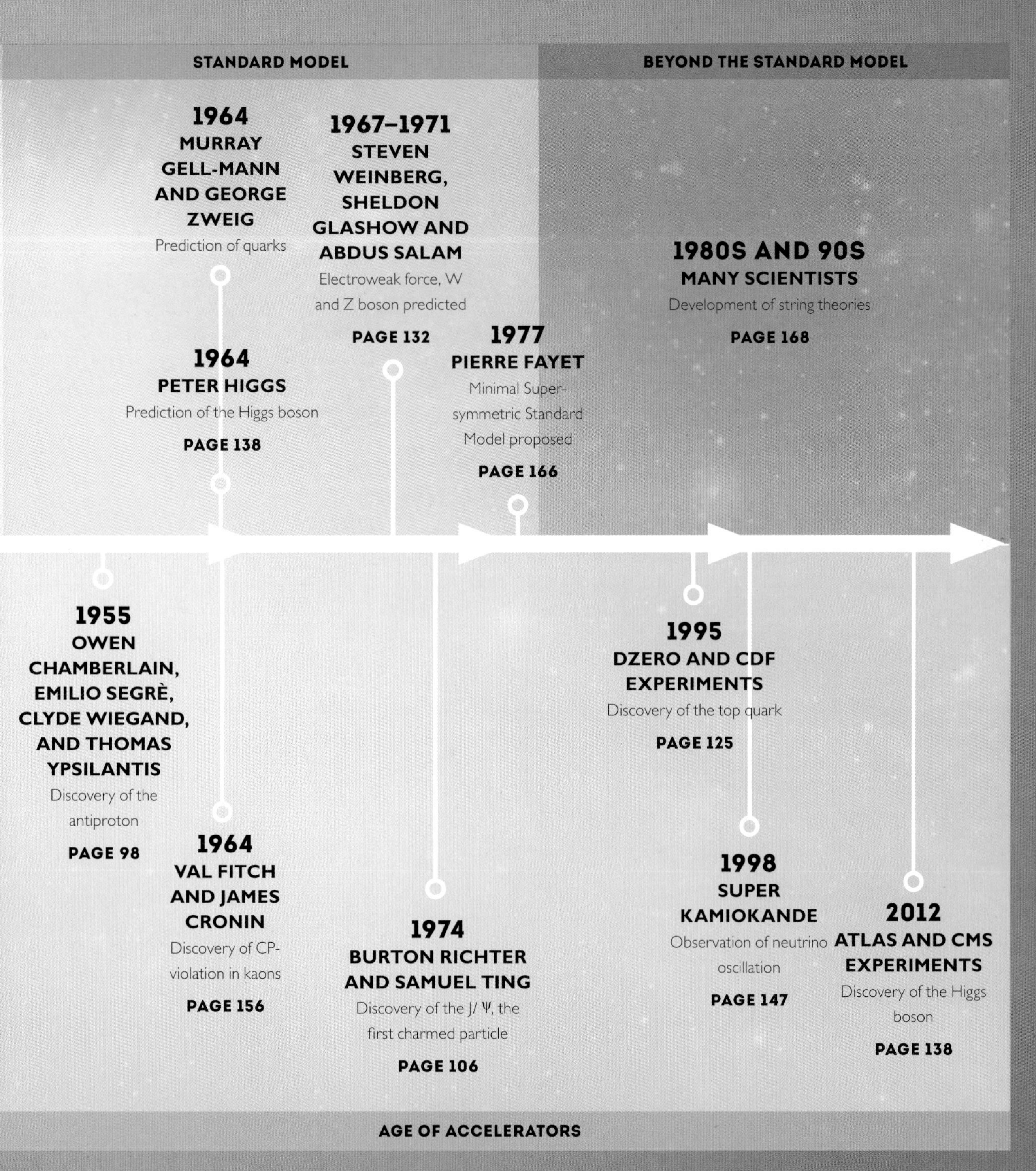
STANDARD MODEL
BEYOND THE STANDARD MODEL
1964
MURRAY GELL-MANN AND GEORGE ZWEIG
Prediction of quarks
1967–1971
STEVEN WEINBERG, SHELDON GLASHOW AND ABDUS SALAM
Electroweak force, W and Z boson predicted
PAGE 132
1980S AND 90S
MANY SCIENTISTS
Development of string theories
PAGE 168
1964
PETER HIGGS
Prediction of the Higgs boson
PAGE 138
1977
PIERRE FAYET
Minimal Super-symmetric Standard Model proposed
PAGE 166
1955
OWEN CHAMBERLAIN, EMILIO SEGRÈ, CLYDE WIEGAND, AND THOMAS YPSILANTIS
Discovery of the antiproton
PAGE 98
1964
VAL FITCH AND JAMES CRONIN
Discovery of CP-violation in kaons
PAGE 156
1974
BURTON RICHTER AND SAMUEL TING
Discovery of the J/ Ψ, the first charmed particle
PAGE 106
1995
DZERO AND CDF EXPERIMENTS
Discovery of the top quark
PAGE 125
1998
SUPER KAMIOKANDE
Observation of neutrino oscillation
PAGE 147
2012
ATLAS AND CMS EXPERIMENTS
Discovery of the Higgs boson
PAGE 138
AGE OF ACCELERATORS

THIS BOOK

We begin this book, like any set of instructions, with a list of the parts – all of the most basic bricks from which our Universe is created. We also discuss important concepts which we will need later on in the book. There will be a lot of new words and necessary jargon here, so it will be a good point of reference as we continue.

To remind you of the list of parts, they are helpfully printed on the pull-out flaps at the front of the book, and the inside flaps have a quick-look glossary of jargon.

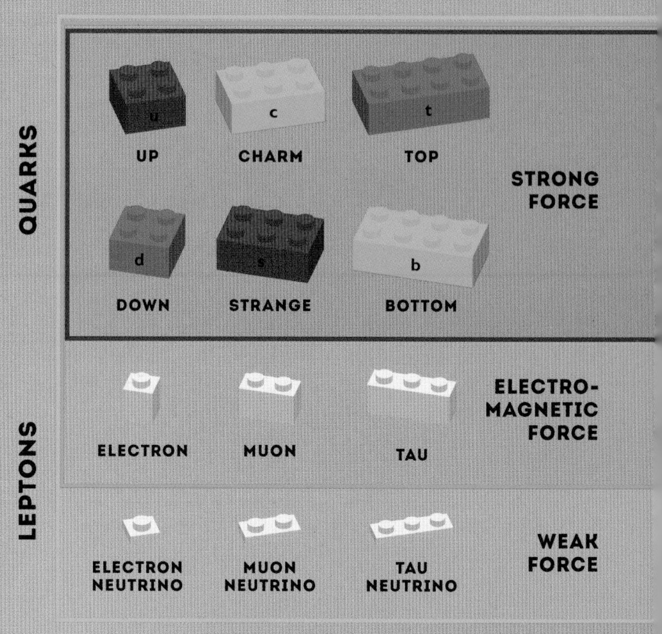

HIGGS FIELD

H

The rest of the book can then focus on the story and we begin at the best place possible, the beginning of time itself – the Big Bang. From here, we chart the almost 14 billion year creation of all normal matter around us, from quarks and electrons to the chemical elements. This story involves only a handful of the particles, the first and least massive generation: up and down quarks; electrons and neutrinos; and antielectrons and antineutrinos. This part of the story covers the first atoms, through the birth and life of stars to their explosive deaths.

The second section of the book discusses the forces of Nature in detail while charting the human story of our ever-growing understanding of particle physics. This story takes us back in time as experiments create conditions ever closer to those not experienced since the Big Bang. There have been thousands of people who have dedicated their lives to advancing human knowledge in this field. While I would like to acknowledge every one I am afraid that it will be just the headline names that get mentioned here. Once we have explored the forces in greater detail you can then use the list of extended construction rules to build an entire zoo of exotic particles of your very own.

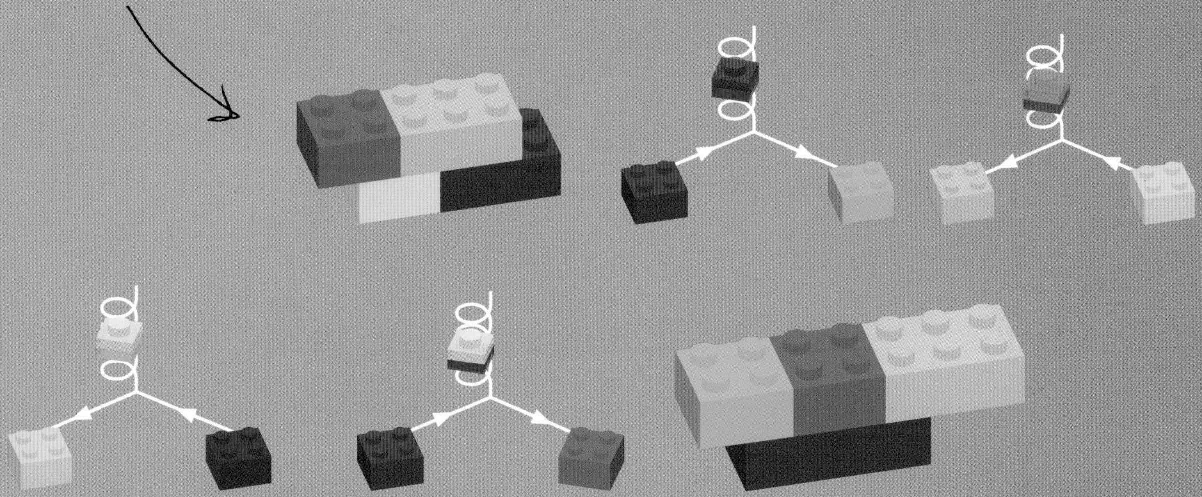

Finally we discuss the future of particle physics – the gaps in our knowledge and some of the theories which hope to fill them. New particles are being introduced as well as new rules for construction. While they might be only theoretical in practice, they can exist in our brick universe.

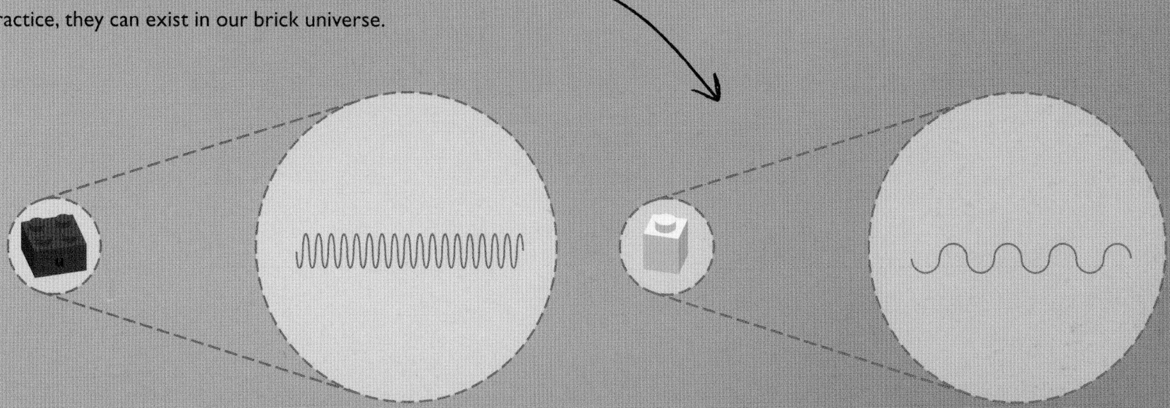

All of the matter around us is built from chemical elements, each made from a unique type of atom

ATOMS

The Greek philosopher Democritus, who lived from 460 to 370 BC, imagined all matter could be divided up into smaller and smaller pieces until eventually there would be a particle which could no longer be divided. This smallest unit of matter he named the atom, from the ancient Greek *atomos* which translates as indivisible. Democritus's atomic idea of matter survived through to the 18th century, a time when chemistry came out of the shadow of alchemy and into the science mainstream. Evidence was building that pure chemical elements reacted together to form new compound chemicals in whole ratios of their mass, for example oxygen and hydrogen reacted in a ratio of 8:1 to form water (note this ratio refers to the total atomic weights not the number of atoms). Englishman John Dalton suggested that this was evidence that chemicals were built from smaller pieces, atoms, and that each chemical element was made from different types of atom. Dalton's new atomic theory proved successful until the end of the 18th century when strange radiation showed that atoms were not as their name suggested - they were divisible to yet smaller parts.

In this chapter we delve inside the atom to the particles from which every atom is made.

OXYGEN + 2 HYDROGEN = WATER

THE PERIODIC TABLE

Toward the end of the 19th century over 100 elements had been discovered, each made from a unique type of atom. Periodically changing patterns, like increasing octaves in music, were used by Dmitri Mendeleev to place these elements into the periodic table. Similarities in behaviour of elements in the same group of the periodic table suggested that they shared some common structure. But if the atoms are, as the Greek word atomos means, uncuttable then how can there be anything smaller? It was not long before things less massive than the lightest atom hydrogen were discovered. These unexpected new 'atoms' seemed to be the reason for the phenomenon of electricity. The discovery of new radiation emitted from atoms confused the subject further and it soon became clear that science was not all wrapped up.

The first row of hydrogen (H) and helium (He) are the elements found in large quantities in the early Universe (pages 42–3) while all elements up to iron (Fe) are made in stars (pages 58–9) and other elements are made in the explosive deaths throes of stars (pages 60–1).

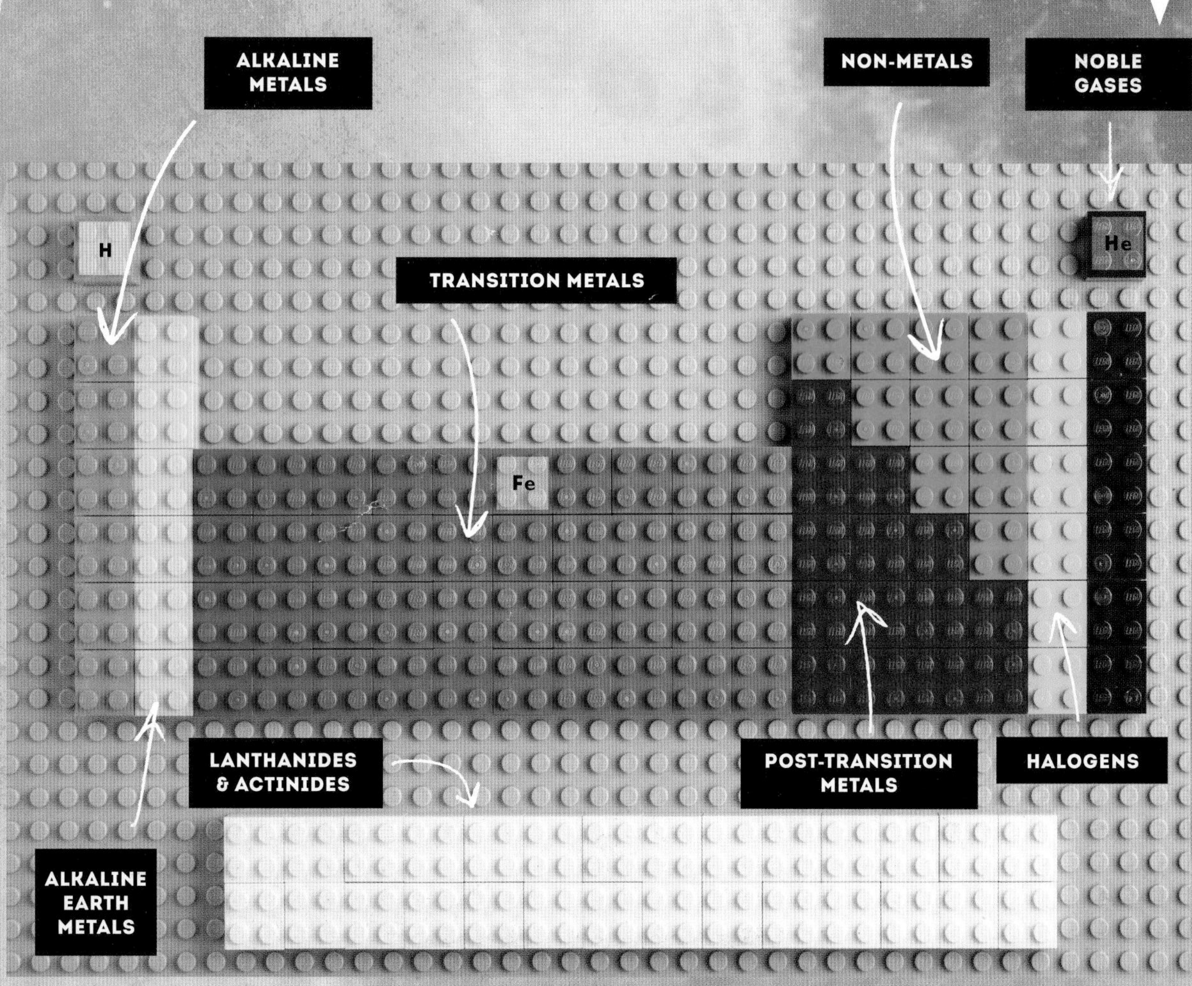

The wide variety of atoms in Nature are all composed from just three smaller, subatomic things

INTO THE ATOM

The work of John Dalton showed us that the world around us is not made from just one type of atom, as the Ancient Greeks thought. The study of electricity and radioactive elements soon showed that Dalton's atoms were also made from other things. The patterns in the modern periodic table only make complete sense once we understand that atoms are constructed from smaller building blocks. Atoms owe their atomic weights to some of these smaller particles, while their reactivity comes from others.

ATOMIC NUCLEUS
This lies at the centre of every atom and is made from the nucleons – protons and neutrons. The nucleus is the source of the radioactivity seen coming from some chemical elements.

WHAT'S IN A NAME?

Protons get their name from the Greek for first. Neutrons are so called as they are electrically neutral. Electrons are named such as they are the atom of electricity.

ELECTRONS

These lie in clouds surrounding a central nucleus. Electrons participate in and dictate the chemical reactions of an element. It is the number of electrons within an atom which determines the periodic nature of chemical reactions.

NUCLEONS

These are not the end of the journey, They are made from yet more fundamental building blocks called quarks. Each proton and neutron is made from a combination of up quarks and down quarks.

UP QUARK

DOWN QUARK

QUARKS

u
d
d
NEUTRON

d
d
u
PROTON

NUCLEONS

STRINGS

Through all of the stress testing they have been subjected to it seems that electrons and quarks are where the journey ends. They do not seem to be constructed from anything else – they seem truly fundamental. However, there are theories which suggest they may well be made of yet smaller things called strings (see pages 168–9).

NEUTRINOS AND LEPTONS

Alongside quarks and electrons, neutrinos complete the range of fundamental particles. Neutrinos are produced in beta decay when a neutron decays into a proton and an electron, which is always accompanied by a neutrino. The electron and neutrino form a group which are called leptons. Being lighter than the protons and neutrons known at the time, they take their name from the Greek *leptós* which means small, thin or delicate.

There are heavier copies of each fundamental particle, identical in every way apart from their mass

MASSIVE COPYCATS

The up quark and down quark, which make up the proton and neutron, and the electron and associated neutrino, are just one generation of fundamental particle; the lightest in mass. For an unknown reason Nature also contains two sets of heavier doppelgangers, shown here as generation 2 and 3, the larger bricks. These particles are only created at very high energies, such as those generated in particle accelerators and which existed early on in the history of our Universe.

		GENERATION 1	GENERATION 2	GENERATION 3	*The twelve particles below are given the name fermions, after the Italian theoretical particle physicist extraordinaire Enrico Fermi.*
FERMIONS	QUARKS	u UP	c CHARM	t TOP	**Up-type quarks** Charge +2/3 The charm quark and the top quark are both heavier than the up. All three particles have positive electric charge but only two-thirds the size of the electron's.
		d DOWN	s STRANGE	b BOTTOM	**Down-type quarks** Charge -1/3 The strange quark and the bottom quark are heavier than the down. All have a negative electric charge which is just one-third of the size of the electron's charge.
	LEPTONS	e^- ELECTRON	μ^- MUON	τ^- TAU	**Charged leptons** Charge + 2/3 The muon and the tau have identical properties to the electron except they are heavier. All three have an electric charge, traditionally labelled negative, and so are known as the charged leptons.
		ν_e ELECTRON NEUTRINO	ν_μ MUON NEUTRINO	ν_τ TAU NEUTRINO	**Neutrinos** Charge 0 All three charged leptons have an associated neutrino particle – the electron, muon and tau neutrino. While their masses have not yet been measured experimentally, we do know that they cannot be massless like photons (see pages 146–7).

Enrico Fermi

The heavier versions of quarks and leptons build exotic forms of matter which are unstable – the heavier the particles they make, the quicker they decay into lighter forms of matter. Eventually all matter decays to form up quarks, down quarks, electrons or neutrinos which is why this first generation of particles make up 99.9% of all visible matter in the Universe.

BUILDING BLOCKS

While searches continue, these twelve particles arranged in three generations seem to be the true fundamental building blocks of all matter, whether it is everyday matter like the chemical elements or exotic matter like that created in particle accelerators and the early Universe.

Fermion particles are special because they stack just like bricks inside atoms, with no one brick occupying the same space. In this way the electrons in atoms take different energies which directly lead to different chemical reactivities of the elements. Protons and neutrons, made from quarks, also fill different energy levels which lead to atomic nuclei being more or less stable. This is how these particles combine together to make atoms with different properties and varying size.

HYDROGEN

HELIUM

LITHIUM

BERYLLIUM

BORON

There are twelve antimatter particles which are mirror opposites of the matter particles

WHAT'S THE ANTIMATTER?

It has also been predicted and observed that the twelve fermion matter particles each have a mirror opposite, a particle where everything about their behaviour is reversed. These are the fundamental particles of antimatter.

Where an electron would be said to have a negative electric charge its antimatter version the antielectron (also called the positron) has a positive electric charge. Flipping the properties of the particles like this is done in a special mirror known as charge inversion (C) which changes a particle to an antiparticle by flipping its charge. Particles and antiparticles interact with each other in totally opposite ways.

In these diagrams, the arrow and $\mathcal{C}$ indicates a charge inversion.

An electron can be changed to an antielectron using a charge inversion.

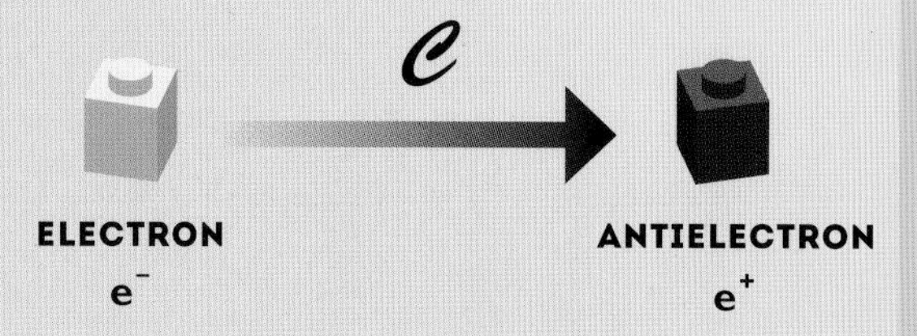

A tau neutrino changes to a tau antineutrino using a charge inversion.

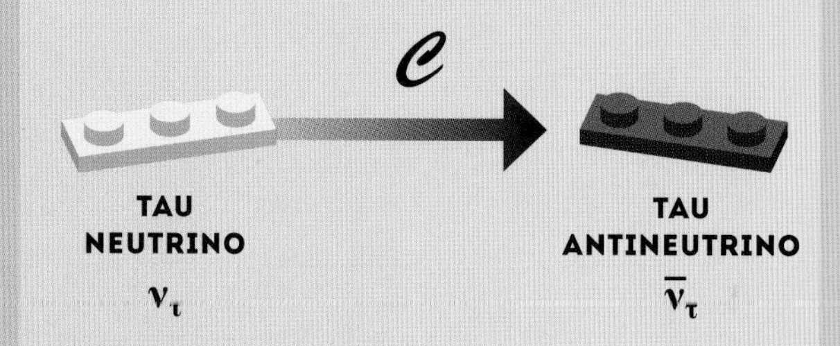

A down quark changes to an antidown quark by charge inversion.

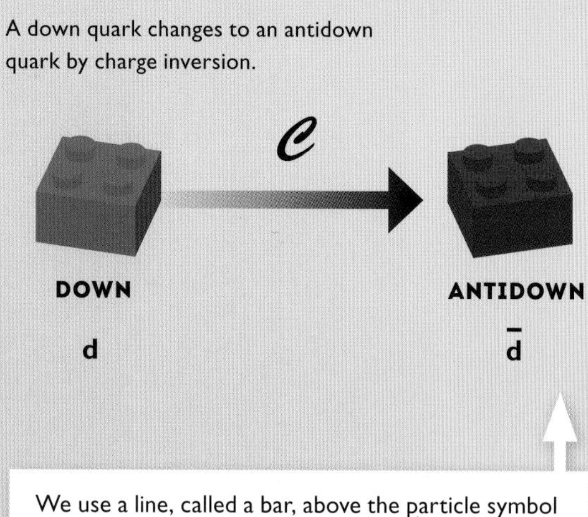

We use a line, called a bar, above the particle symbol to show it is the antiparticle.

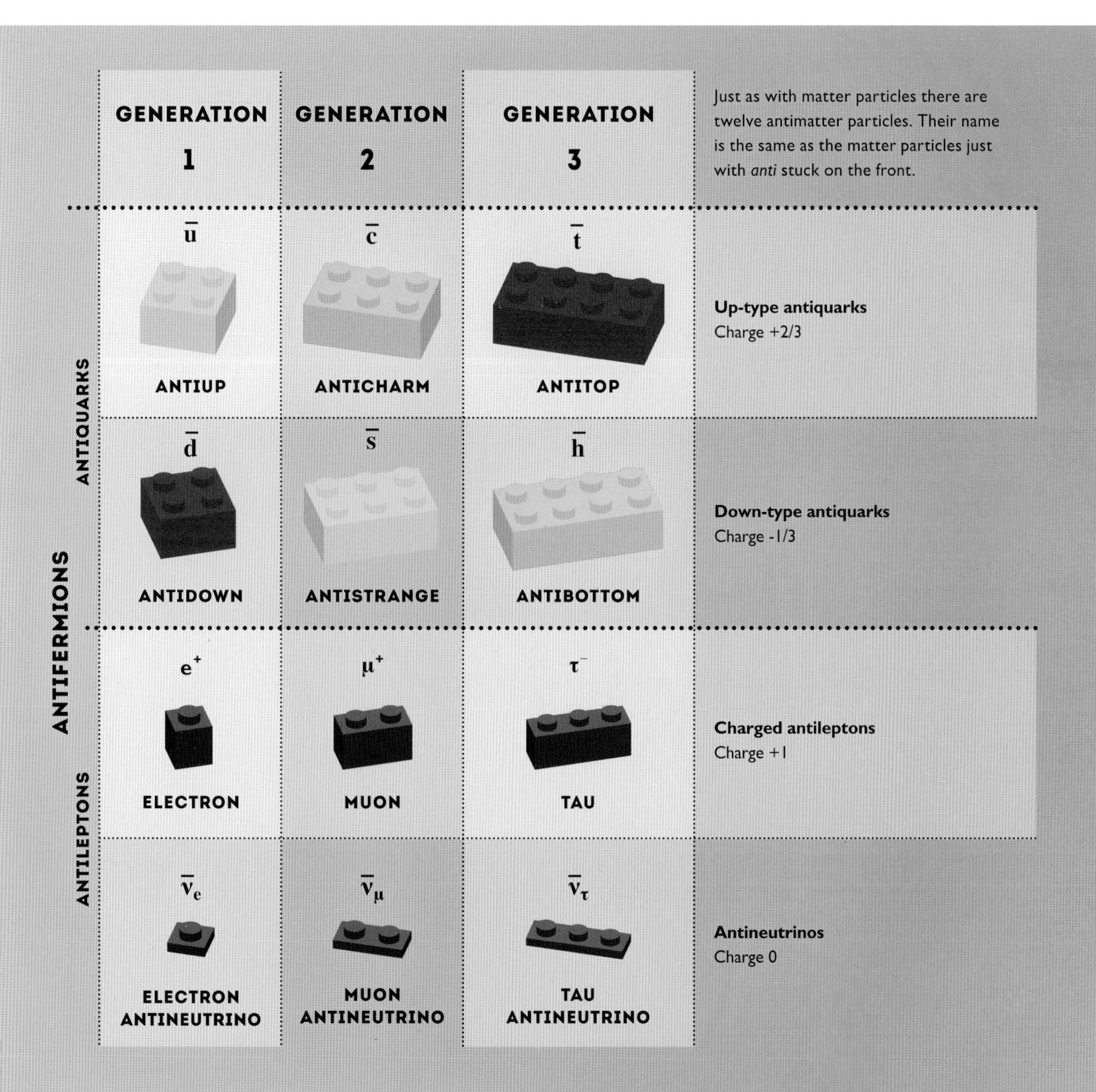

Today antimatter is created in high-energy locations such as particle accelerators (see pages 98–9), and the particles colliding with our atmosphere (see pages 94–5). Antimatter would have existed naturally in abundance in the hot early Universe but it has long since been obliterated. When matter and antimatter meet one another they annihilate each other to form pure energy – usually in the form of light.

The forces dictate how particles interact and combine to form atoms, elements and chemicals

FORCES

Now we have a set of building blocks we need some building instructions and these are dictated by the forces of Nature. Rather than step-by-step explicit instructions, these forces impose guidelines and restrictions as to how fundamental particles interact and combine with one another. Nature seems dictated by the influence of four fundamental forces. Two you may be familiar with as they have an infinite range of influence – the forces of gravity and electromagnetism. The other two will be less familiar as they are confined to the nucleus of an atom – the weak and strong nuclear forces. The influence of each force is exchanged between fermion particles by a separate class of particles called bosons, named after the Bengali theoretical physicist Satyendra Nath Bose.

GRAVITY

While gravity will be the most familiar force as it is firmly fixing you to the surface of this planet, it is the weakest fundamental force. This weakness and the tiny masses of particles mean that gravity does not really affect the lives of particles as it does our own. Gravity is ignored when considering the interaction of particles.

GRAVITONS

The graviton is the boson particle predicted to exchange the, always attractive, gravitational force between objects with mass. Gravitons are not part of the Standard Model and remain theoretical, having not been observed experimentally. The recent discovery of gravity waves however, provides indirect evidence of gravitons – where there is a force field there are usually bosons.

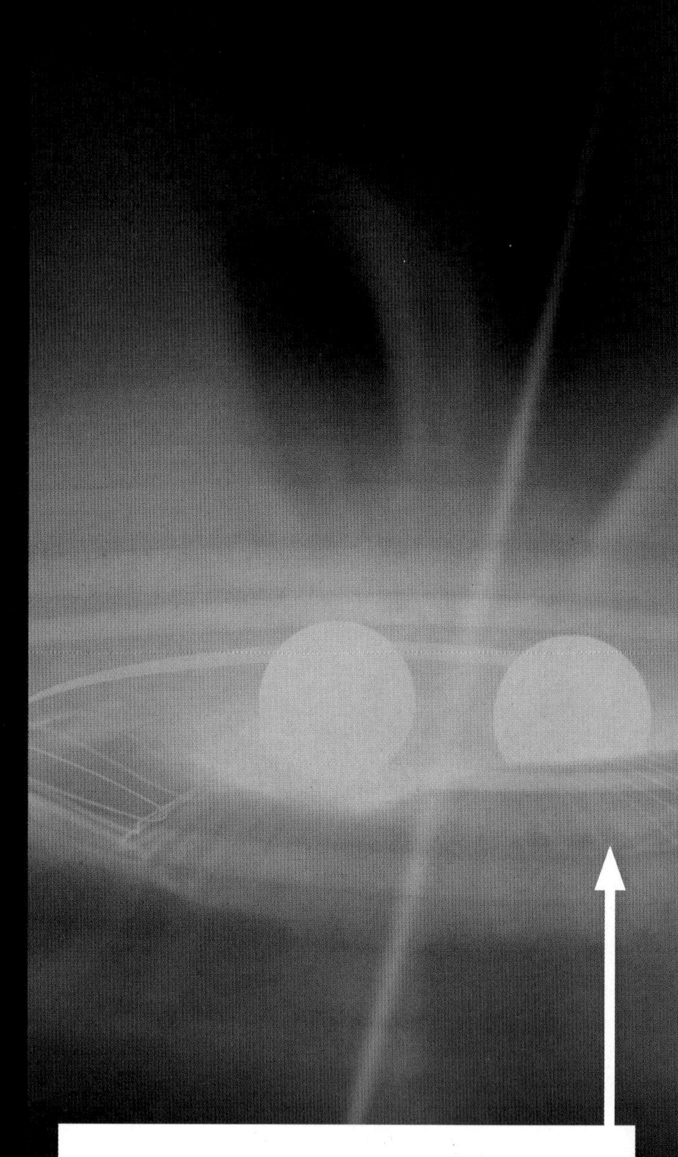

Gravity waves might, like all other forces, be made from the concerted behaviour of boson exchange particles: gravitons.

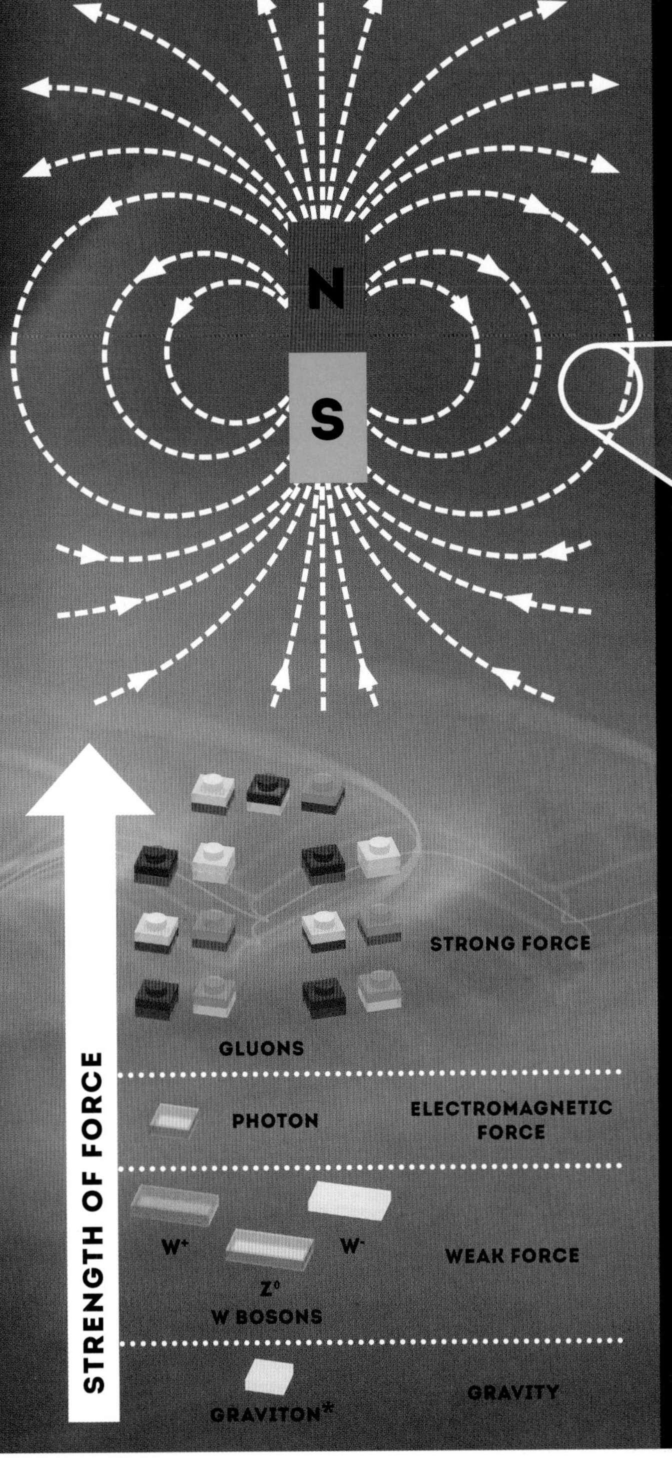

FIELDS

When looking at the force exchange between just a few fundamental particles it makes sense to think of the exchange of bosons. Everyday interactions with the forces, however, involve many trillions of particles exchanging many more bosons – the maths of which would be horrific. In these real-world scenarios we instead talk of force fields, such as the magnetic field, which is in essence an average influence of all of these bosons. The electromagnetic fields arise from the movement of particles called photons.

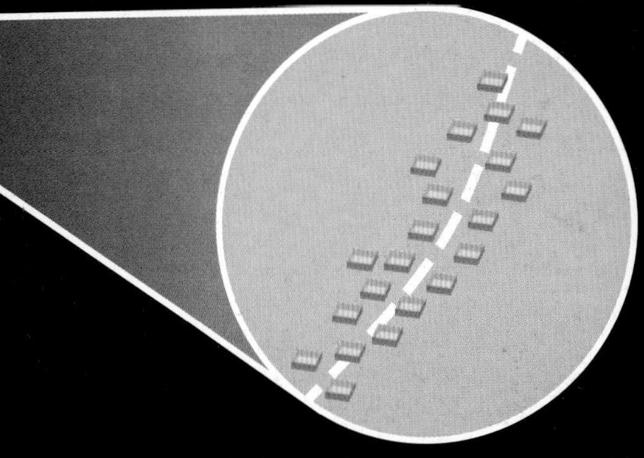

BOSONS

The influence of each force is exchanged between quarks and leptons by a separate set of particles called bosons.

With the exception of gluons (the exchange particles of the strong force (see page 27)), bosons do not connect together like fermions to build more complex matter. Instead they are just messengers which exchange information and energy between particles.

On the left we can see the four forces in order of strength along with associated bosons.

* *This particle has been hypothesised but never observed experimentally*

The electromagnetic force is electricity and magnetism combined and is the reason why everything feels solid

ELECTROMAGNETISM

We experience the electromagnetic force regularly in everyday life, whether it is a static shock from an escalator's electric field, or the magnetic fields which keep pictures of loved ones on our fridge door. Experiments in the early 19th century by Michael Faraday and the mathematical brilliance of James Clerk Maxwell showed that these two phenomena arise from the same force of electromagnetism. Maxwell also showed us that the force is exchanged by an electromagnetic radiation. Maxwell's ideas were developed further by physicists in the 1960s to account for the quantum behaviour of particles, forming the theory of QED (quantum electrodynamics). In QED the electromagnetic force is exchanged between fermions by particles of light called photons. Photons are particles of light with no mass and so travel at the maximum speed possible in the Universe, aptly called the speed of light.

All electrically charged particles experience the electromagnetic force and can exchange, absorb or emit, photons. All electrically charged fermion particles in our analogy are full-height plastic bricks.

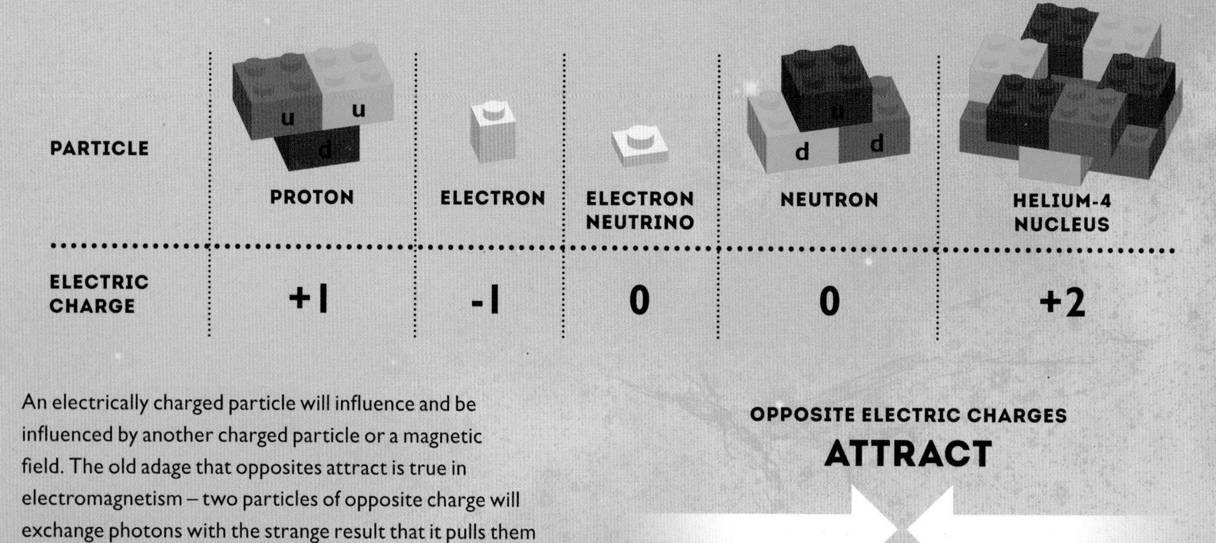

An electrically charged particle will influence and be influenced by another charged particle or a magnetic field. The old adage that opposites attract is true in electromagnetism – two particles of opposite charge will exchange photons with the strange result that it pulls them together (this can also be seen between the North and South poles of a magnet).

OPPOSITE ELECTRIC CHARGES

ATTRACT

u u d

PHOTON

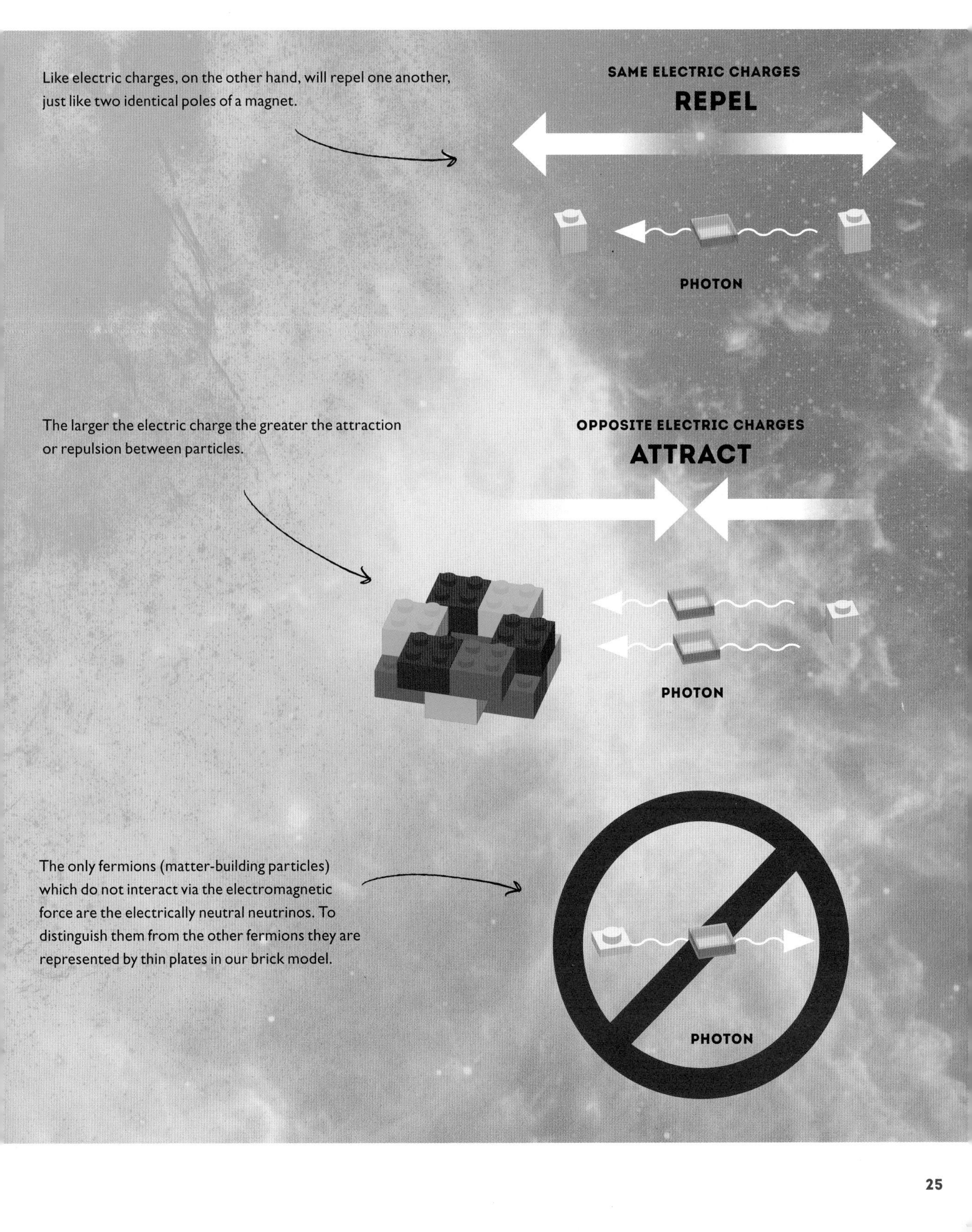

Like electric charges, on the other hand, will repel one another, just like two identical poles of a magnet.

The larger the electric charge the greater the attraction or repulsion between particles.

The only fermions (matter-building particles) which do not interact via the electromagnetic force are the electrically neutral neutrinos. To distinguish them from the other fermions they are represented by thin plates in our brick model.

Without the strong force to bind positive electrically charged protons, no atoms could exist

STRONG FORCE

The strong force is indeed the strongest force of all, but its influence is confined to a tiny region around a particle, a space no bigger than a proton. In an atom the primary job of the strong force is to overcome the electromagnetic repulsion between protons in the nucleus. It is aptly named strong because if this force were not stronger than the electromagnetic force then the atomic nucleus would not exist, and the protons would just push apart. Meson particles exchange the strong force (see pages 92–3) which pulls protons and neutrons together in the atomic nucleus.

The strong force is attractive between all protons and neutrons, pulling them together to form a nucleus.

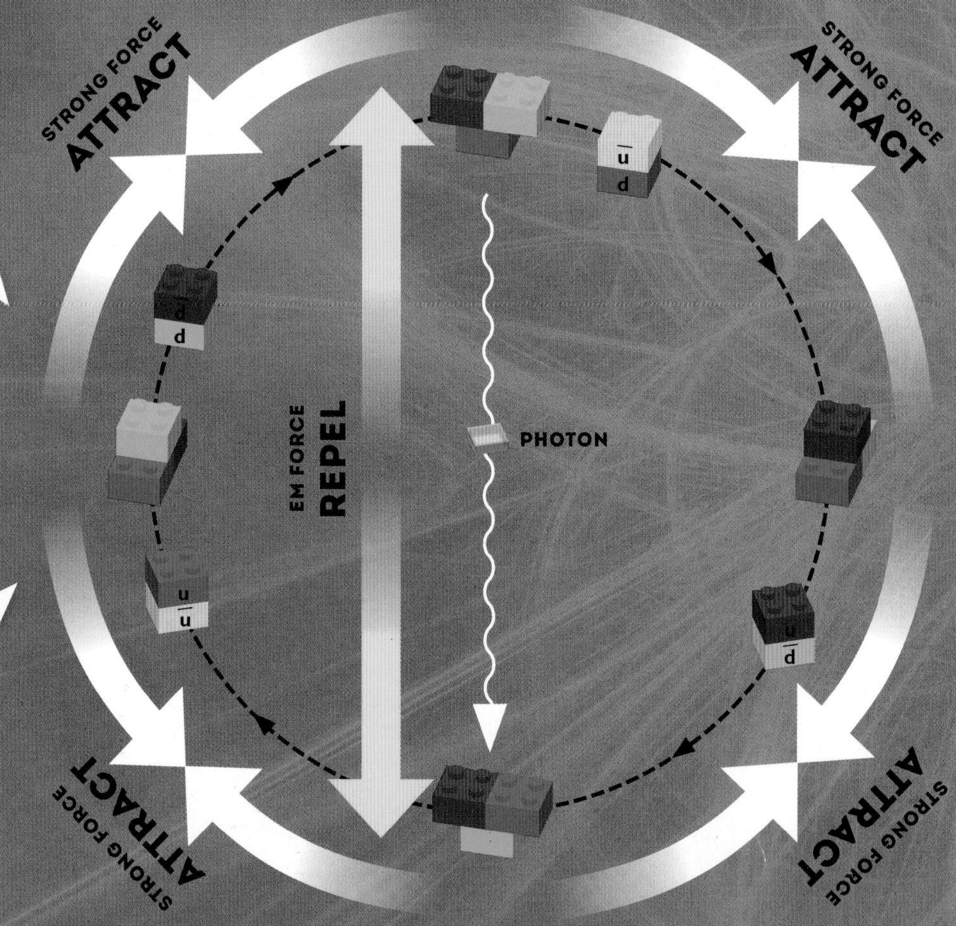

The electromagnetic force between two same, positively charged, protons is repulsive – pushing against the strong force and fighting against the formation of a nucleus.

STRONG FORCE
ATTRACT

STRONG FORCE
ATTRACT

STRONG FORCE
ATTRACT

The strong force is also responsible for gluing quarks into groups to form composite particles like the protons, neutrons and many exotic forms of matter. The strong force is exchanged between quarks by boson particles imaginatively called gluons.

The strong force only influences the lives of quarks and antiquarks which are the only particles which can therefore absorb or emit a gluon. This is shown in our model by quarks and gluons being colourful, while leptons (which are not influenced by the strong force) are black or white.

LEPTONS

Leptons do not feel the strong force at all and so never interact with gluons. So in our brick model the leptons are not colourful. Leptons are just white and antileptons are black.

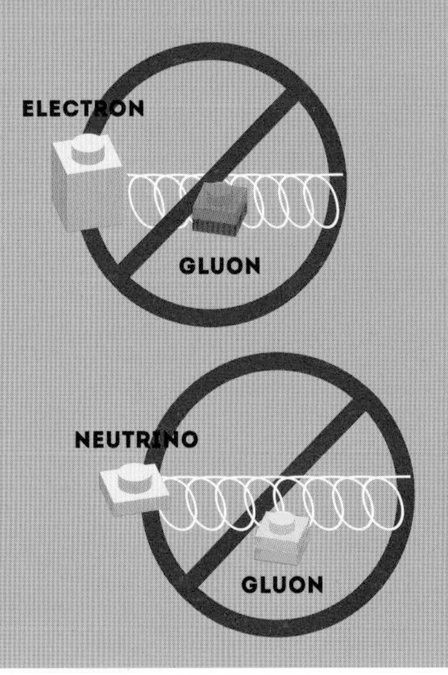

The weak force is felt only across the nucleus of an atom and transforms fundamental particles

WEAK FORCE

The fourth and final force is the weak force, so called because now, at the current temperature of our Universe, it is not as strong as either the electromagnetic or strong force. The weak force does not build structures like the other forces but is instead responsible for bizarre interactions which change particles. It can change quarks from up-like (up, charm or top) to down-like (down, strange or bottom) and vice versa, or change a charged lepton into a neutrino. The weak force is the only force to be felt by *all* fermion particles (the electromagnetic force is not felt by electrically neutral neutrinos and the strong force is only felt by quarks).

DECAY

The weak force's ability to change particles is particularly useful when heavy particles or atomic nuclei want to change into lighter particles, a process called decay. It is also useful when lighter atomic nuclei have the right conditions to become heavier such as the process of fusion which occurs in the centre of every star (more on this in chapter 2).

The change of particle happens thanks to the exchange of electrically charged W-bosons – W^+ and W^-. Look back at page 23 to see which bosons relate to each force.

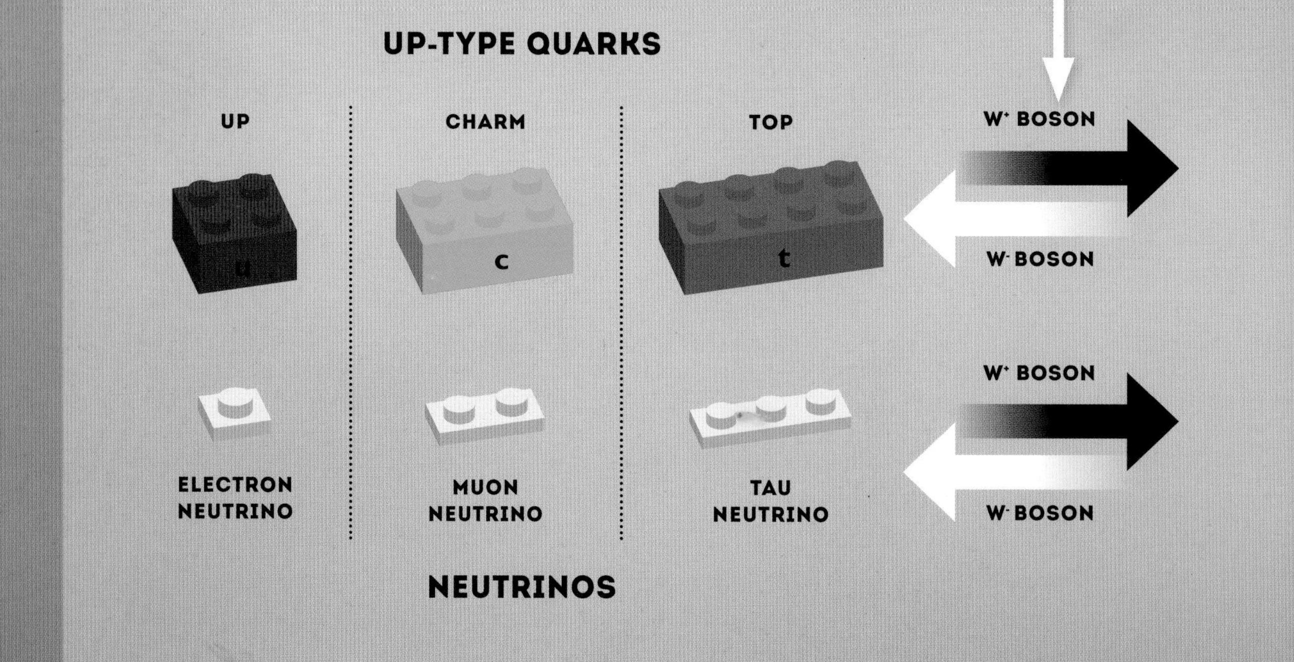

A NEUTRAL PARTY

Alongside the electrically charged W weak bosons, there exists an electrically neutral weak force boson particle – the Z-boson, Z^0. Essentially a photon with a mass, the Z^0 exchanges energy but leaves everything else about a particle unchanged.

FERMIONS AND BOSONS

In this plastic brick analogy all of the fermions have studs on top because they can connect together to form more complex matter. For example, quarks combine to produce particles like protons and neutrons. Electrons combine with protons and neutrons to make atoms.

The weak force bosons and the photon are smooth tiles because these particles cannot combine to make more complex things. They cannot form matter themselves, they can only exchange forces.

Gluons on the other hand can interact with each other and can theoretically form exotic matter called glueballs (see pages 112–3) which is why they do have studs in the model.

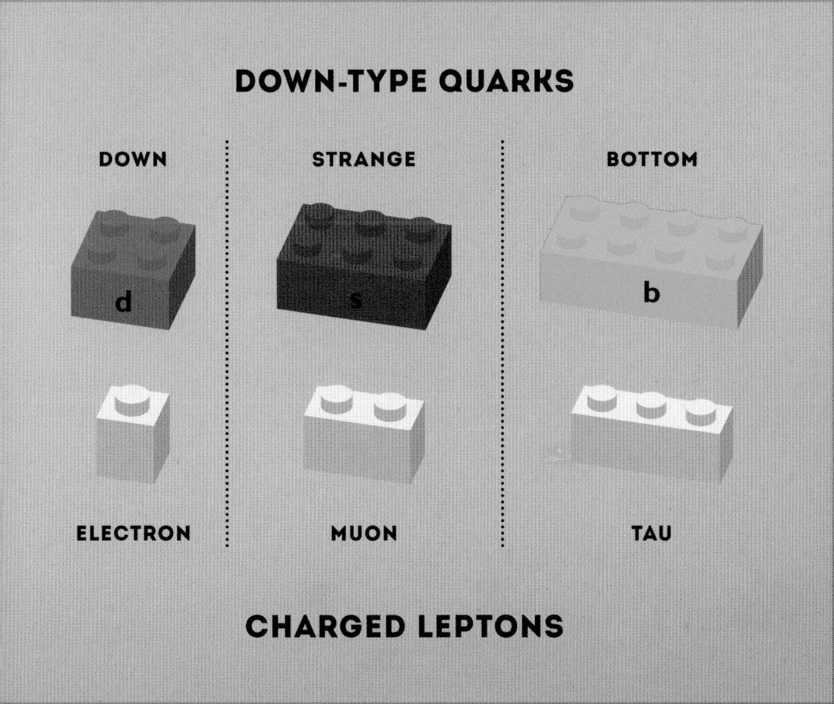

DOWN-TYPE QUARKS

DOWN | STRANGE | BOTTOM

d | s | b

ELECTRON | MUON | TAU

CHARGED LEPTONS

LEPTONS AND NEUTRINOS

Leptons are matter-building fermion particles that do not feel the strong force. Some, such as the electron, muon and tau, have an electric charge and so interact with the electromagnetic force. Neutrinos have no electric charge and so can only interact with other particles via the weak force. Each electrically charged lepton has an associated neutrino which can be found paired together in any weak force interaction. The W weak force boson transforms electrically charged leptons into their neutrino partner and vice versa.

The Higgs Boson slows particles down, which appears to give them mass

HIGGS BOSON

Then of course there is the Higgs boson. Predicted in 1965 and seen with confidence in 2012 it is the final piece of the jigsaw puzzle that is the Standard Model of particle physics. The Higgs boson is the particle associated with the Higgs field. The Higgs field is a different type of force, not one that allows particles to interact with one another, but one which only allows a particle to interact with the Higgs boson, or field, itself. In the same way that a magnetic field interacts with other magnetic objects nearby, the Higgs field interacts with all fundamental particles which have a mass (all fermions and the weak force bosons).

The Higgs field, through the exchange of Higgs bosons, slows down the fundamental particles as if they were travelling through treacle. It prevents particles from ever being able to reach the fastest speed possible – the speed of light. The slowing down acts like our everyday understanding of mass, where a greater force is needed to increase the speed of an object with a large mass compared to an object with a small mass.

The Higgs boson does not interact with all particles in the same way, interacting with some more than others. The more the Higgs boson interacts the greater it slows a particle down and the more massive the particle is observed to be.

A conceptual illustration of a Higgs field.

TOP QUARK

This is the most massive fundamental particle and interacts with the Higgs boson the most.

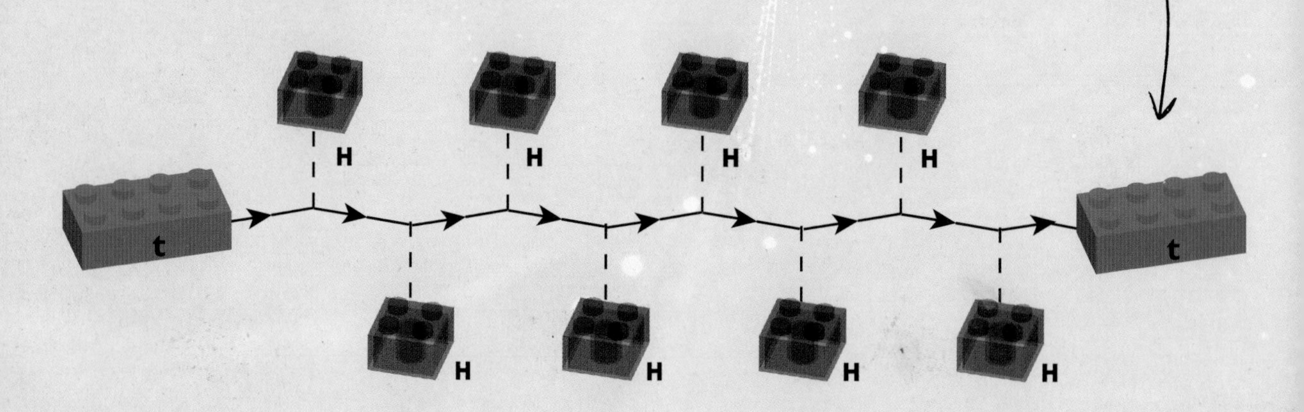

LIGHT ELECTRONS

These interact far less with the Higgs boson and can therefore be accelerated to high speeds more easily than the top quark.

NEUTRINOS

These have tiny and as yet undetermined mass, so out of all the fermions they interact the least with the Higgs field.

HIGGS BOSON

The Higgs boson also has a mass because it can interact with itself, slowing itself down much like a commuter dawdling along the street playing a game on their phone.

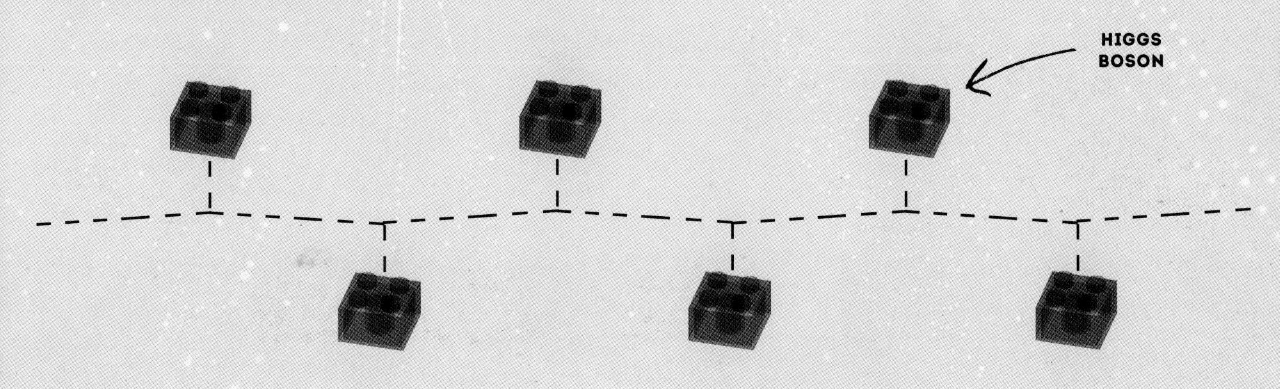

We use new units based on the electron-volt, eV, to measure the world of tiny particles

ENERGY AND MASS

In our daily lives we measure the energy of things in units of Joules, after English physicist and brewer of beer James Prescott Joule. Everyday things are made from trillions of atoms and even more subatomic particles. If we picked just one from this multitude its energy would be tiny. The energies of individual particles are instead measured in units of electron-volts, eV with 1 eV equalling a tiny 1.6×10^{-19}, or 0.00000000000000000016, Joules. Rather than write lots of zeros and quote energy in Joules we convert the unit of energy, much like a currency exchange between different monies. One electron-volt is the energy an electron would receive when accelerated by an electric potential difference of 1 Volt, see page 66 on particle accelerators for more.

The vast amount of energy unleashed in an atomic bomb explosion comes from the conversion of tiny amounts of mass into energy in a fraction of a second.

TEMPERATURE

It is impossible to measure the energy of trillions and trillions of particles, Instead on these human scales we use temperature which is effectively the average kinetic (movement) energy of all particles in a material. In this way the temperature of any matter directly relates to the energy of the particles it is made from. The kinetic energy, stored in the movement of particles, at room temperature is around 0.025 eV. The energy of particles in modern particle accelerators run into the trillions (tera, T) of electron-volts.

MASS

Once again our everyday kilogram measure of mass is impractical when speaking of particles. While an adult human being will weigh in at around 70 kilograms, an electron has a mass of just 9.1×10^{-31}, or 0.00000000000000000000000000000091 kilograms. Again a conversion is made to keep numbers sensible.

This conversion requires us to use Einstein's most famous equation, telling of the interchange between energy and the mass. If you could unleash it, even the smallest amount of mass would release huge amounts of energy as the exchange rate is the speed of light squared $9 \times 10^{16}\,m^2/s^2$. Rearranging the formula we see that energy divided by the speed of light squared equals mass. So we choose the unit of particle mass to reflect the choice of energy unit, eV, by measuring particle masses in units of electron-volts per speed of light squared, or eV/c^2.

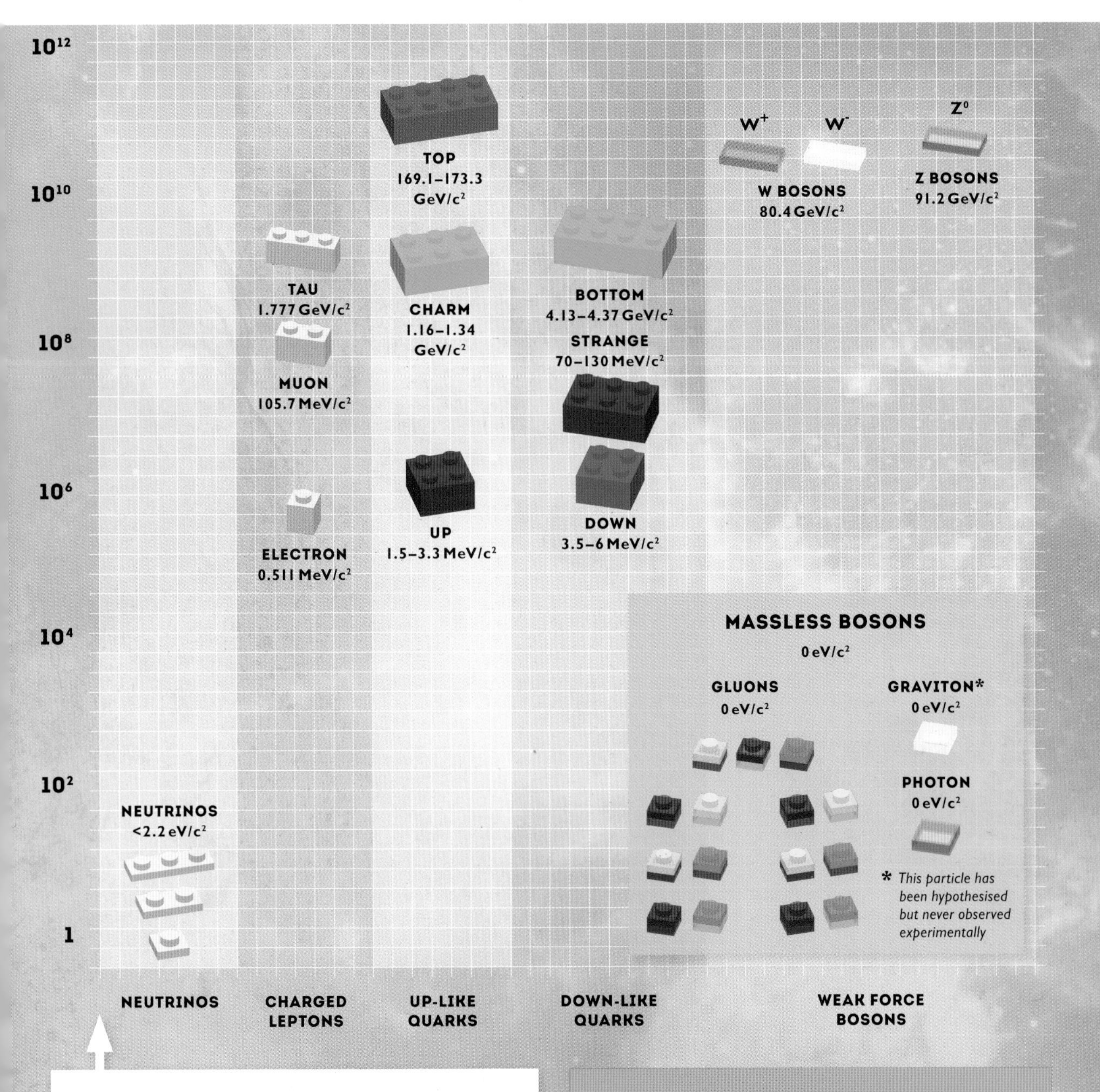

There is a massive factor of 100 billion difference between the mass of the lightest neutrino and that of the heaviest top quark. The mass of each fundamental particle must be measured through experiment and added by hand into the mathematics of the Standard Model.

SCALE PREFIXES

An electron has a mass of 500,000 eV/c², which we can simplify in terms of the powers of ten using scale prefixes.

500,000 eV/c² = 500 keV/c² (k = kilo = ×1000)
= 0.5 MeV/c² (M = mega = ×1,000,000)
= 0.0005 GeV (G = giga = ×1,000,000,000)

It is thought that our Universe began its life as pure energy which expanded from an infinitely small point

THE BIG BANG

About 13.8 billion years ago something happened in the void. Where there was nothing, all of the space, time and energy we see around us unfurled from an infinitely small point. In the first moment it expanded rapidly, faster than the speed of light, a period known as inflation. The high concentration of energy in this early Universe made the temperature extremely hot. Our current understanding of Nature can take us back to 10^{-43} seconds after this Big Bang event, but cannot explain how or why it occurred. Before this time the Universe becomes too high in temperature, dense in energy and small in size for current science to describe.

Shortly after this energy horizon, however, we understand that the Universe was a bubbling soup of particles exchanging blows as they careered into one another at near light speed. These particles most likely destroyed each other once more to form energy in a cycle of creation and destruction.

As the Universe continued to expand it stretched space and time with it. The more it expanded and energy spread out, the more the Universe cooled down; much like the expanding gas expelled from a spray can cools the contents of the can down. As it cooled, the soup of particles cooled also, slowing them down. At different times particles found themselves unable to simply collide with and bounce off other particles. Forces that acted between the particles started to dictate their behaviour.

This chapter covers the early years of the Universe from the turbulent Big Bang to a more placid place where stars shone their first light into a dark cosmos.

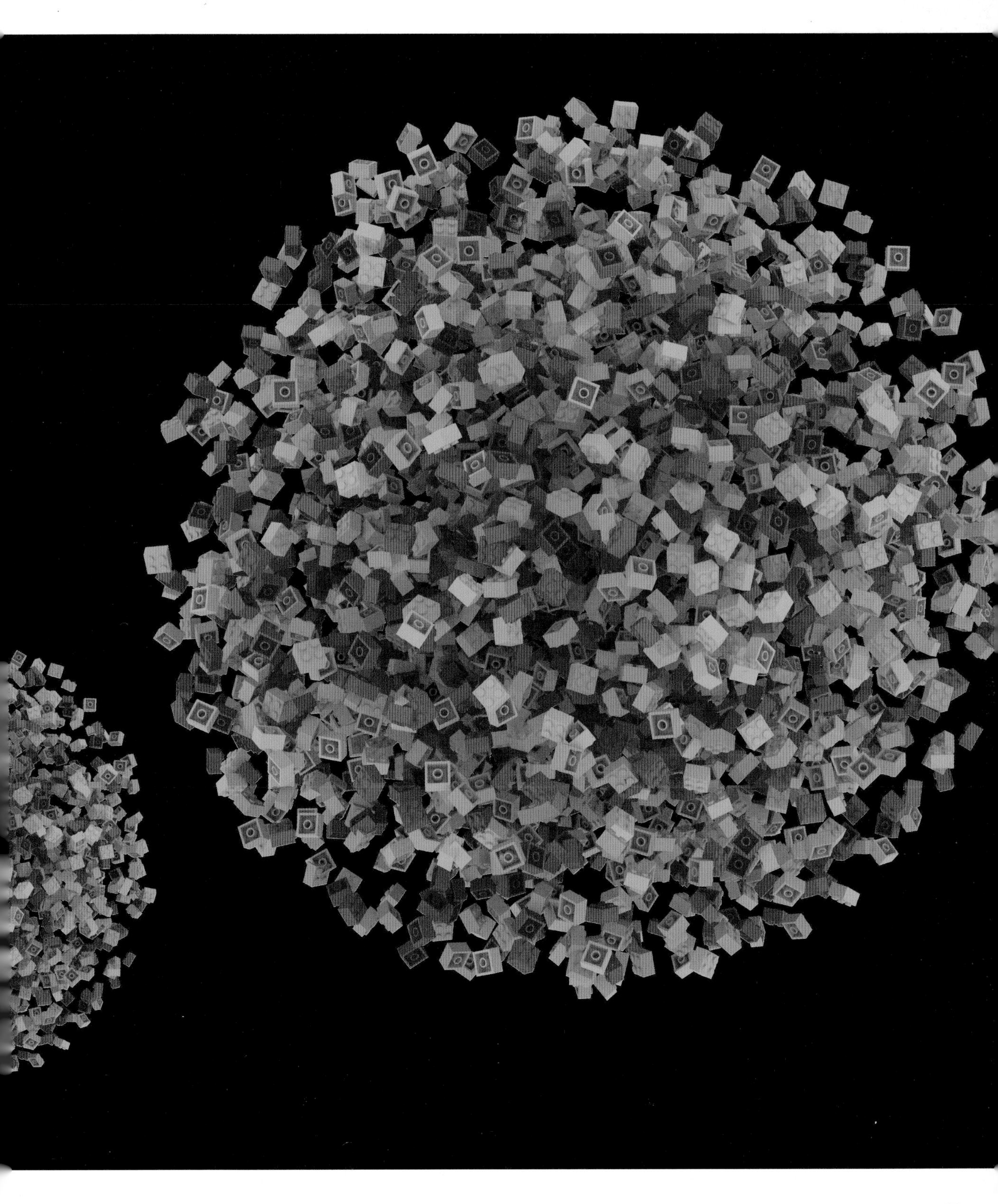

The history of the Universe is one of expansion and cooling

A BRIEF HISTORY

This diagram provides an overview of the next few pages and is a summary of the times in our Universe's history which are of particular importance to a particle physicist. Here is a key to understand the sections on each stage:

Seconds – Time after the Big Bang that each force took hold of particles. From far less than a second through to the age of the Universe of about 13.8 billion years.

K – Temperature of the Universe at key times when forces took hold of the particles. The temperatures are in units of degrees Kelvin rather than the more familiar degrees Celsius or Fahrenheit. You can see this changes from octillions of degrees soon after the Big Bang to just 2.7 K, the temperature of deep space, today.

Each degree Kelvin is equivalent to a degree Celsius, with the only difference being that 0 on the Kelvin scale starts at -273.15 °C. This is the coldest possible temperature possible, known as absolute zero.

GEV – Average energy of fundamental particles in the Universe at each time in giga electron-volts (one billion electron-volts). It ranges from hundreds of billions of electron-volts of energy in the first fraction of a second, to a sluggish few ten thousandths of this energy today.

Energy is the particle physicist's time machine. Huge amounts of energy is given to particles creating conditions which existed soon after the Universe was created. The energy of protons in the Large Hadron Collider at CERN gives physicists a glimpse of what things were like less than 100 billionths of a second after the Big Bang.

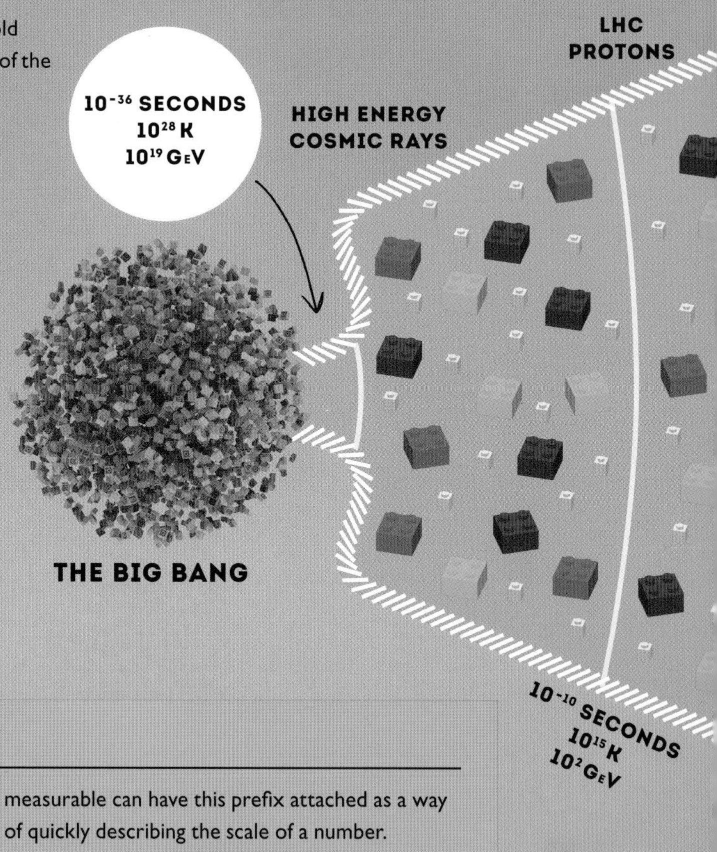

UNDERSTANDING MEASUREMENTS

We will occasionally use scientific prefixes in this book due to the variety in the scale of sizes and energies. You may be most familiar with large prefixes from the world of computing; megabyte is one million, or 10^6, bytes, a gigabyte is one billion, or 10^9, bytes. You may also be familiar with small prefixes from measurements of size; a millimeter is one thousandth, or 10^{-3}, of a meter. Anything measurable can have this prefix attached as a way of quickly describing the scale of a number.

If you are not familiar with standard notation, 10^{-4} written out in full is 0.0001 and 10^2 is 100. The number is moved to the *right* of the decimal point by the *negative* number of places or to the *left* of the decimal point the *positive* number of places.

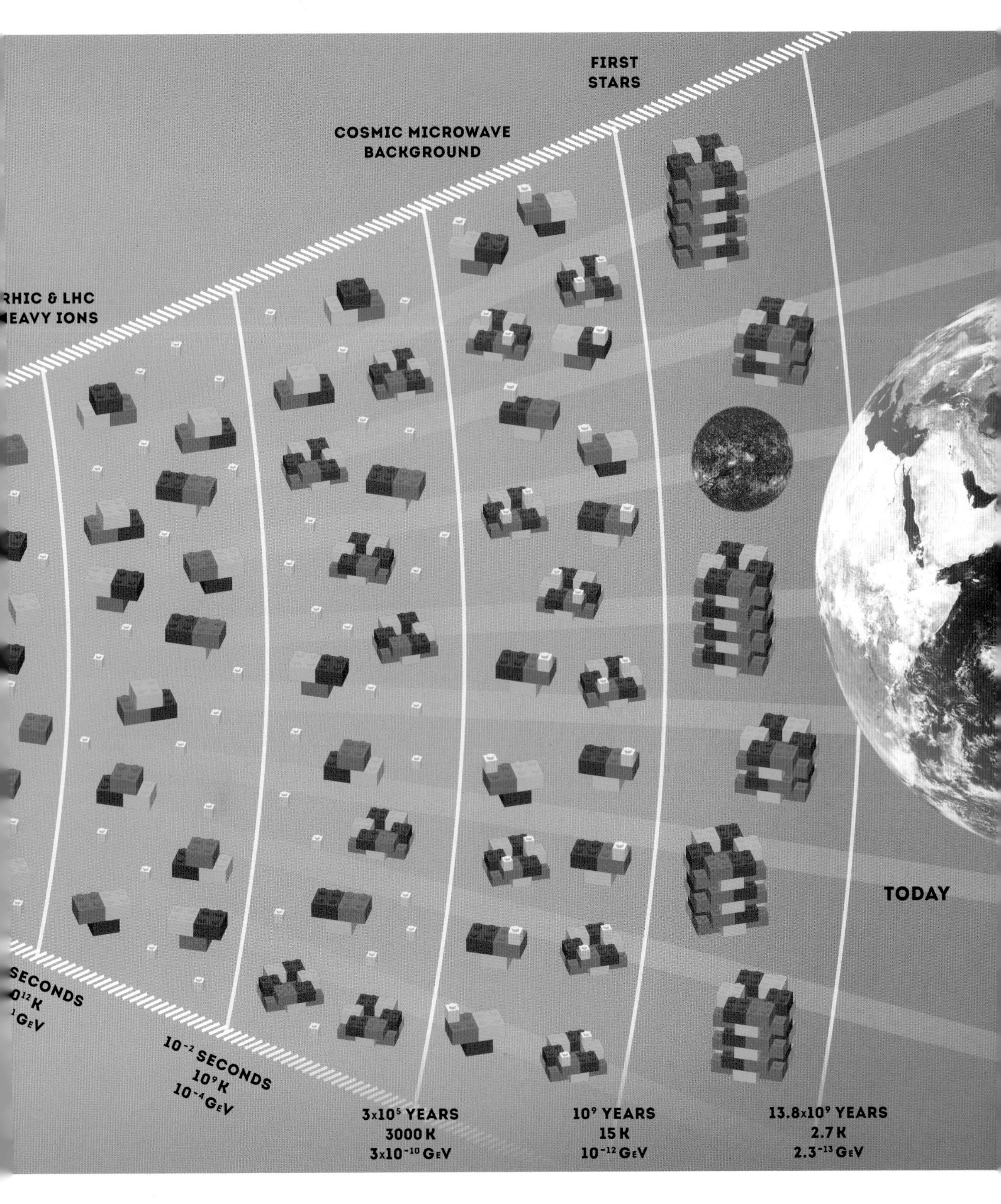
FIRST
STARS
COSMIC MICROWAVE
BACKGROUND
RHIC & LHC
EAVY IONS
SECONDS
K
GeV
10^-2 SECONDS
10^9 K
10^-4 GeV
3x10^5 YEARS
3000 K
3x10^-10 GeV
10^9 YEARS
15 K
10^-12 GeV
13.8x10^9 YEARS
2.7 K
2.3^-13 GeV
TODAY

After the first fraction of a second, quarks could no longer overcome the strong force and combined to form protons

THE FIRST SECOND

At around 10^{-10} (or 0.0000000001) of a second, matter particles dominated in number over their opposite number – antimatter. The exact cause of this is something particle physicists are working on today (see CP violation, pages 154–5). At the massive temperatures of about 1,000,000,000,000,000 K the matter particles were moving so fast that they just bounced off one another.

At around 10^{-4} second old the Universe had cooled to a point where one group of the matter particles called quarks could no longer just bounce off each other. Instead the strong nuclear force (see chapter 4) forced quarks to bind together into groups of three.

Two types of quark, the up quark and the down quark combine in two different ways to form the first composite particles; particles made up from multiple fundamental particles.

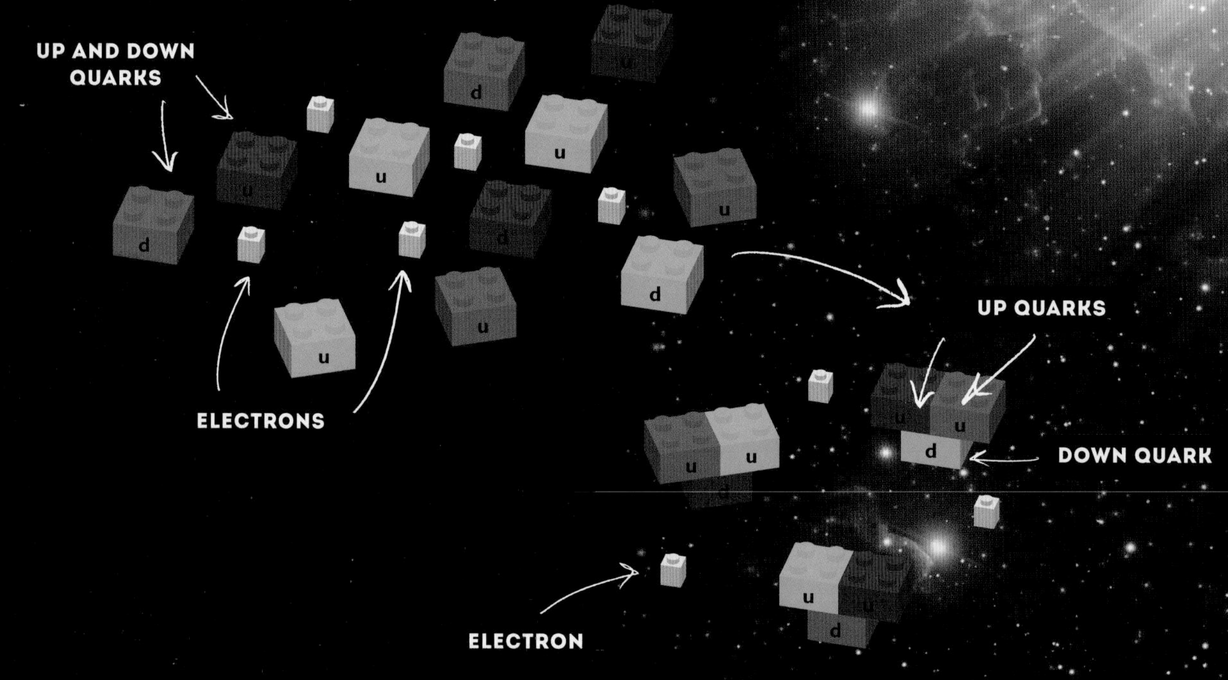

PROTONS & NEUTRONS

To make a proton place two up quarks on top of one down quark, forming a downward-pointing triangle.

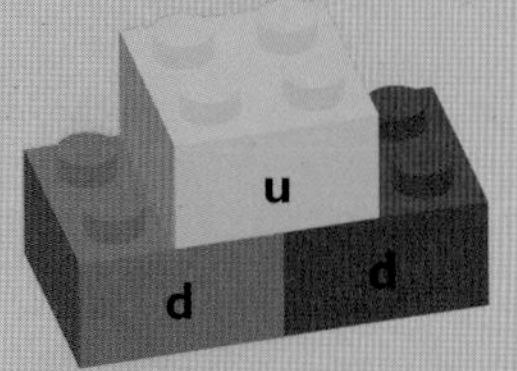

To make a neutron place one up quark on top of two down quarks, making an upward-pointing arrow.

Electron particles just sat by and observed, unaffected by the strong nuclear force.

In each model there must be one red, one green and one blue quark. It does not matter which quark is which colour, as long as all three are present in the proton or neutron you have made.

In the first minutes the weak force transformed protons into neutrons allowing helium nuclei to form

THE FIRST MINUTE

At tens of seconds old the Universe was around one billion degrees Kelvin in temperature. The high temperature and density of matter at this age meant that enough protons collided with one another that some combined, bound together by the strong force that they each feel.

Neutrons cannot survive alone for very long and readily turn into protons. But protons can also morph into neutrons given the right conditions. The weak nuclear force is required for this interchange between neutrons and protons, and is essential for this stage in cosmic evolution.

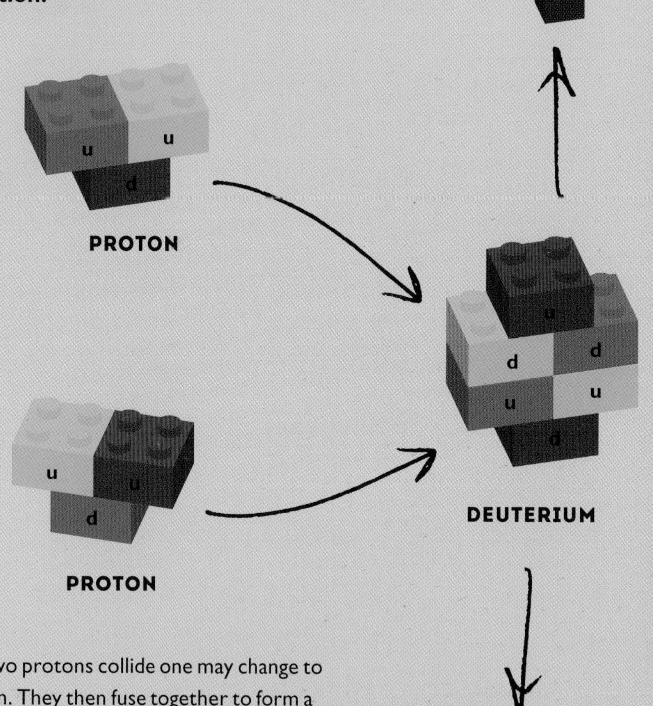

1 When two protons collide one may change to a neutron. They then fuse together to form a deuterium nucleus; to make a deuterium nucleus connect one proton with one neutron. In the change from proton to neutron other particles, the antielectron and electron neutrino, are ejected as by-products.

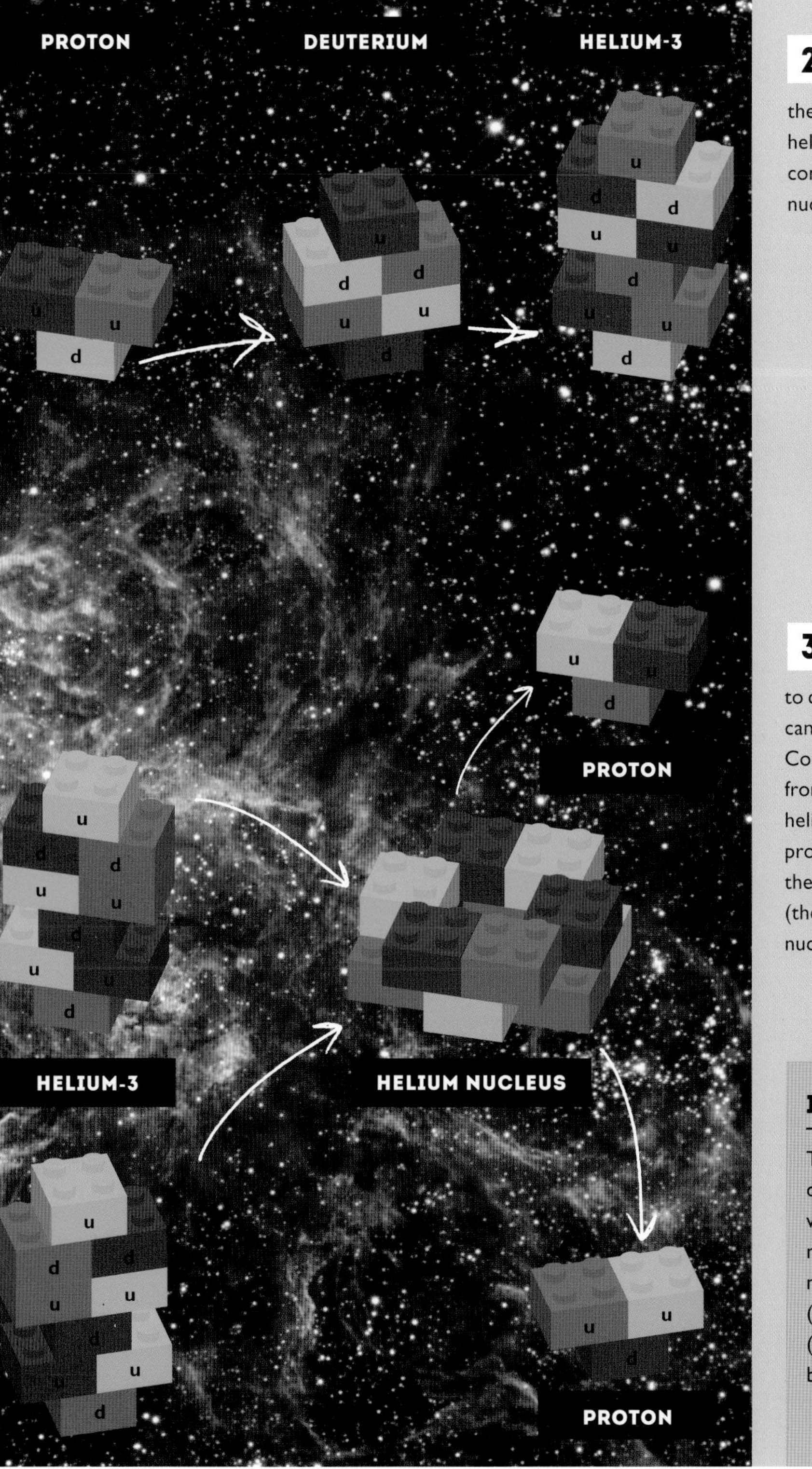

2 If an additional proton collides in a short space of time with the deuterium, they can fuse to form helium-3. To make a helium-3 nucleus connect one proton to a deuterium nucleus.

3 If in their short lifetime two helium-3 nuclei are lucky enough to collide with each other then they can form the stable nucleus helium-4. Combine one proton and one neutron from each helium-3 nucleus to form a helium-4 and discard the additional two protons. Stable helium-4 remains as the second ever type of stable nucleus (the first being the proton, which is the nucleus of a hydrogen atom).

ISOTOPES

The number of protons in a nucleus defines which chemical element it will become. Nuclei with the same number of protons but a different number of neutrons, e.g. helium-3 (2 protons, 1 neutron) and helium-4 (2 protons and 2 neutrons), are both isotopes of helium.

After around 380,000 years electrons could no longer escape their attraction to positive nuclei and combined to form hydrogen and helium

THE FIRST ATOMS

After the creation of some helium-4 the universe remained predominantly a mixture of positive electrically charged nuclei and negative electrically charged electrons. They were attracting and repelling one another by exchanging particles of light called photons. It was these photons that shared the electromagnetic force responsible for interactions between things with an electric charge. This light was trapped, constantly bouncing between nuclei and electrons, being absorbed and emitted. This was the state of the Universe for a long period of time and, because light could not freely spread out into the Universe, it was known as the cosmic dark ages.

PROTON
HELIUM
PHOTON
ELECTRON

HYDROGEN & HELIUM

At around 380,000 years old, at a temperature of just 3000 K, things in the Universe changed. The negative electrically charged electrons no longer had energy enough to resist the electromagnetic attraction of the positive electrically charged nuclei. The two combined to produce the very first atoms of hydrogen and helium.

These new atoms had equal numbers of positive protons in the nucleus and negative electrons attached and so were electrically neutral, with zero overall charge. To form a neutral hydrogen atom, attach one electron to a proton.

To form a neutral helium atom, attach two electrons to a helium-4 nucleus.

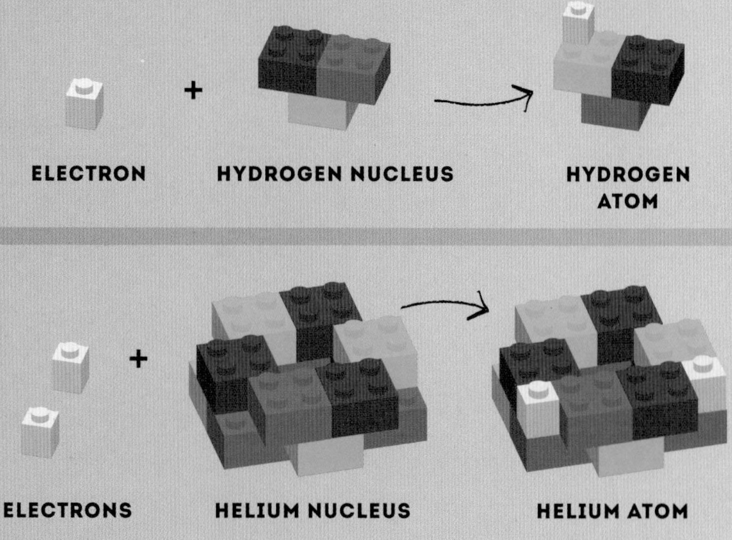

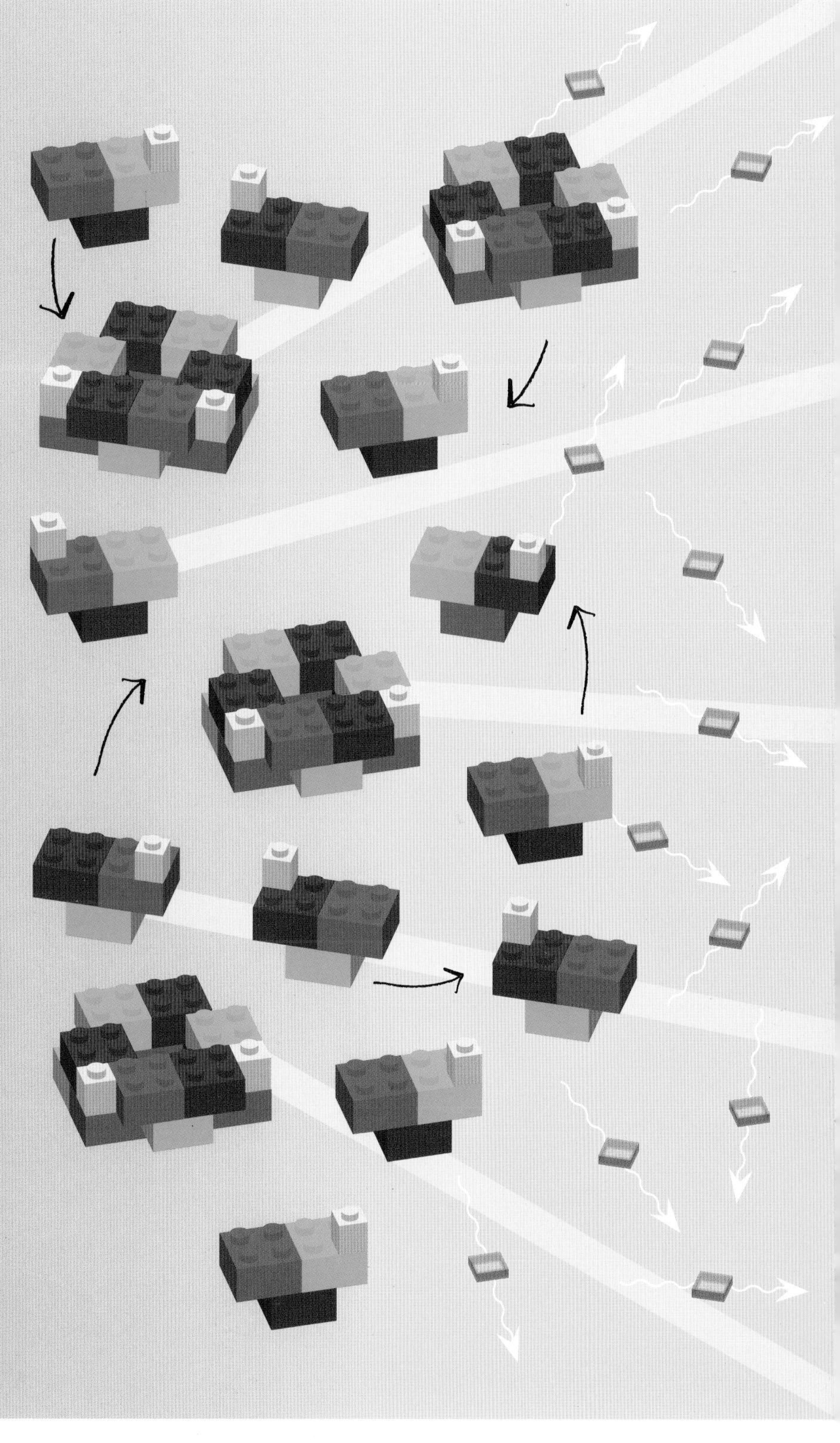

LET THERE BE LIGHT

With few electrically charged things to interact with, the bound photons of light now found themselves free. The light spread out into the cosmos and was stretched from microwave wavelengths to radio waves today as space expanded. Mapping this cosmic microwave background, the Planck satellite and other experiments like it, have provided us with pictures of this oldest light in the Universe. It is the furthest into the past that astronomical telescopes can take us, but, as we will find in the latter chapters of this book, there are ways of going back to the very first second.

PLASMA

Hot mixtures of positive and negative electrically charged particles with interacting photons is known as a plasma. This is viewed by physicists as a fourth state of matter as it is very different from the other three of solid, liquid, or gas.

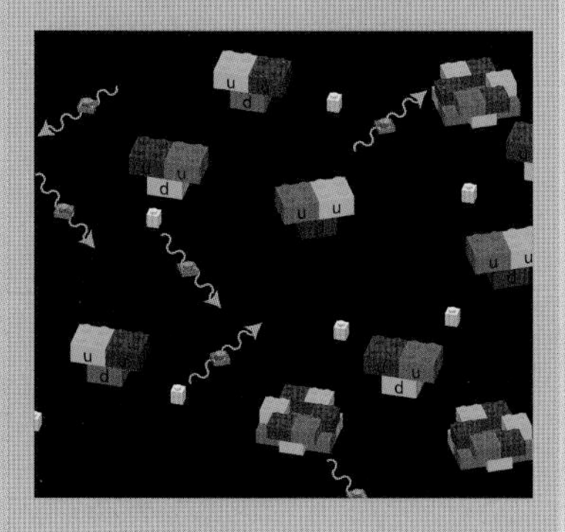

After 100 million years gravity finally takes hold to form the first stars

THE FIRST STARS

We have not yet mentioned the most familiar day-to-day force of gravity. Gravity acts upon the mass that something has. Gravity is extremely weak; our legs can overcome the gravitational attraction between our body and an entire planet when we jump. Electromagnetic force is much stronger and to prove this you need only to use a magnet to lift a large mass of metal easily off the ground. This weakness, combined with the fact that particles have very tiny masses, leads to particle physicists largely ignoring gravity altogether. But, when clouds of light gas get to be light years in size then their combined mass means that gravity finally exerts its influence in the Universe.

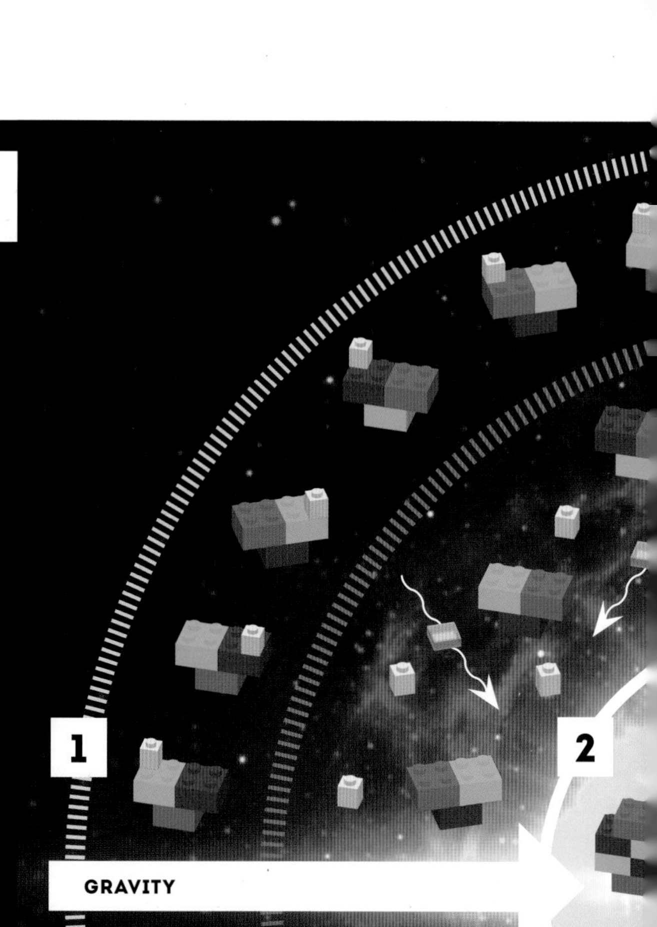

1 Large clouds almost entirely made of hydrogen gas formed around 100 million years after the Big Bang and began to collapse under the force of gravity. As the gas fell inwards it began to heat up. In the centre of the cloud the temperature rose above 3000 K and electrons were again able to free themselves from nuclei to form a plasma.

2 Soon the clouds reached temperatures and densities not seen since the first minute of the Universe's life. In the centre of each cloud a tiny microcosm, called a protostar, recreated the conditions necessary for protons to fuse together. It is here that the first stars were born as fusion ignited and protons combined to form helium.

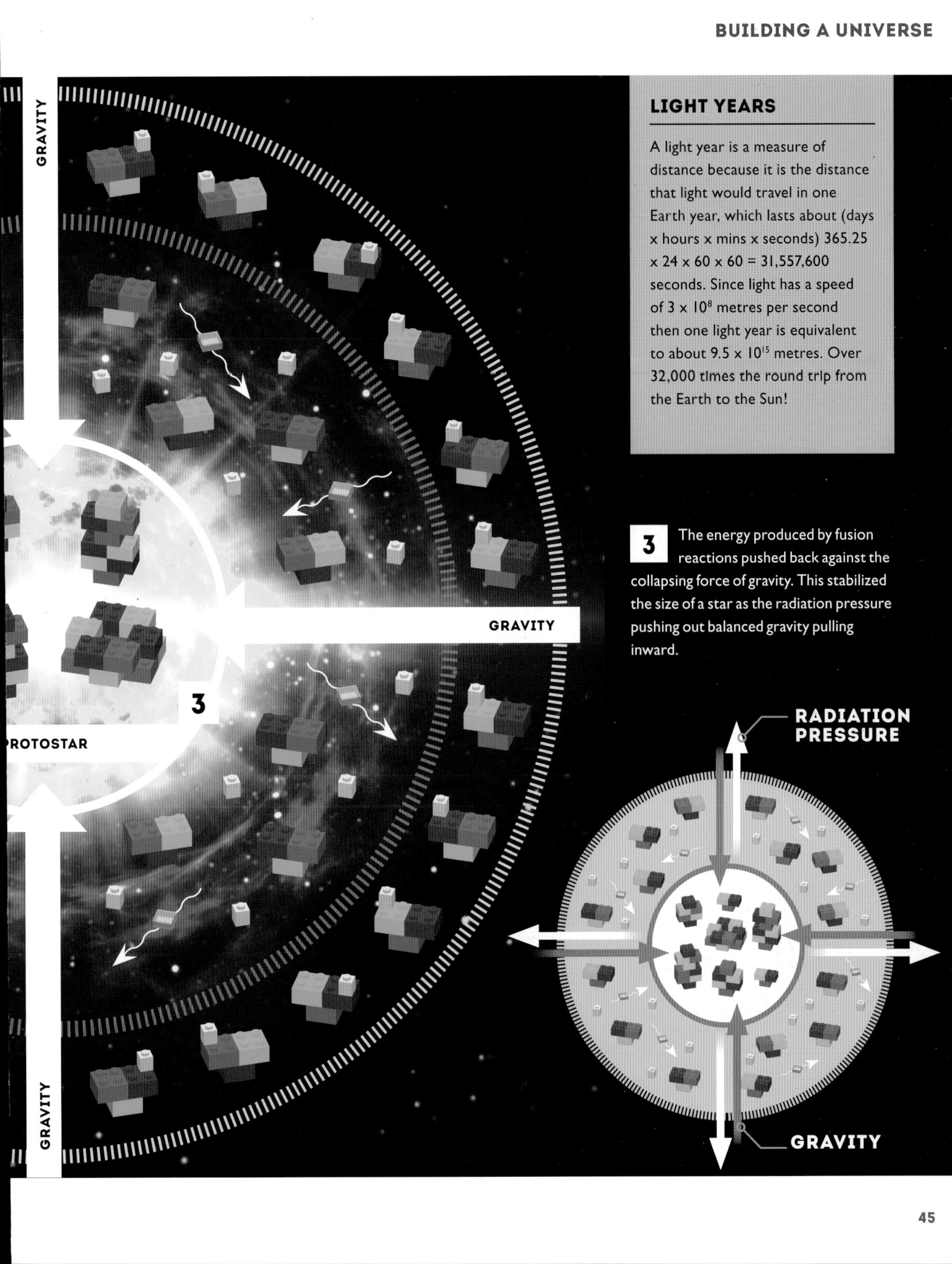

LIGHT YEARS

A light year is a measure of distance because it is the distance that light would travel in one Earth year, which lasts about (days x hours x mins x seconds) 365.25 x 24 x 60 x 60 = 31,557,600 seconds. Since light has a speed of 3×10^8 metres per second then one light year is equivalent to about 9.5×10^{15} metres. Over 32,000 times the round trip from the Earth to the Sun!

3 The energy produced by fusion reactions pushed back against the collapsing force of gravity. This stabilized the size of a star as the radiation pressure pushing out balanced gravity pulling inward.

The high density at the centre of a star results in more collisions between protons, turning hydrogen into helium

THE PROTON–PROTON CHAIN

While it has a similar temperature to the early Universe, the centre of a star is more tightly packed with protons. This increased density results in more protons and nuclei colliding with one another than ever before. An increase in collisions led to the chances of new nuclei being created. The proton–proton chain is an extension of the fusion that occurred in the first minutes. All stars, no matter their size or age, will spend over 90% of their life creating helium from protons through this chain. There are a host of other possible routes that result in the production of helium-4, but here is the most probable one:

1 As we've seen on page 40, two protons can collide to form a deuterium, releasing an antielectron and an electron neutrino. When the deuterium collides with another proton it forms helium-3. Two helium-3 nuclei colliding results in a helium-4 nucleus, with two protons being shed.

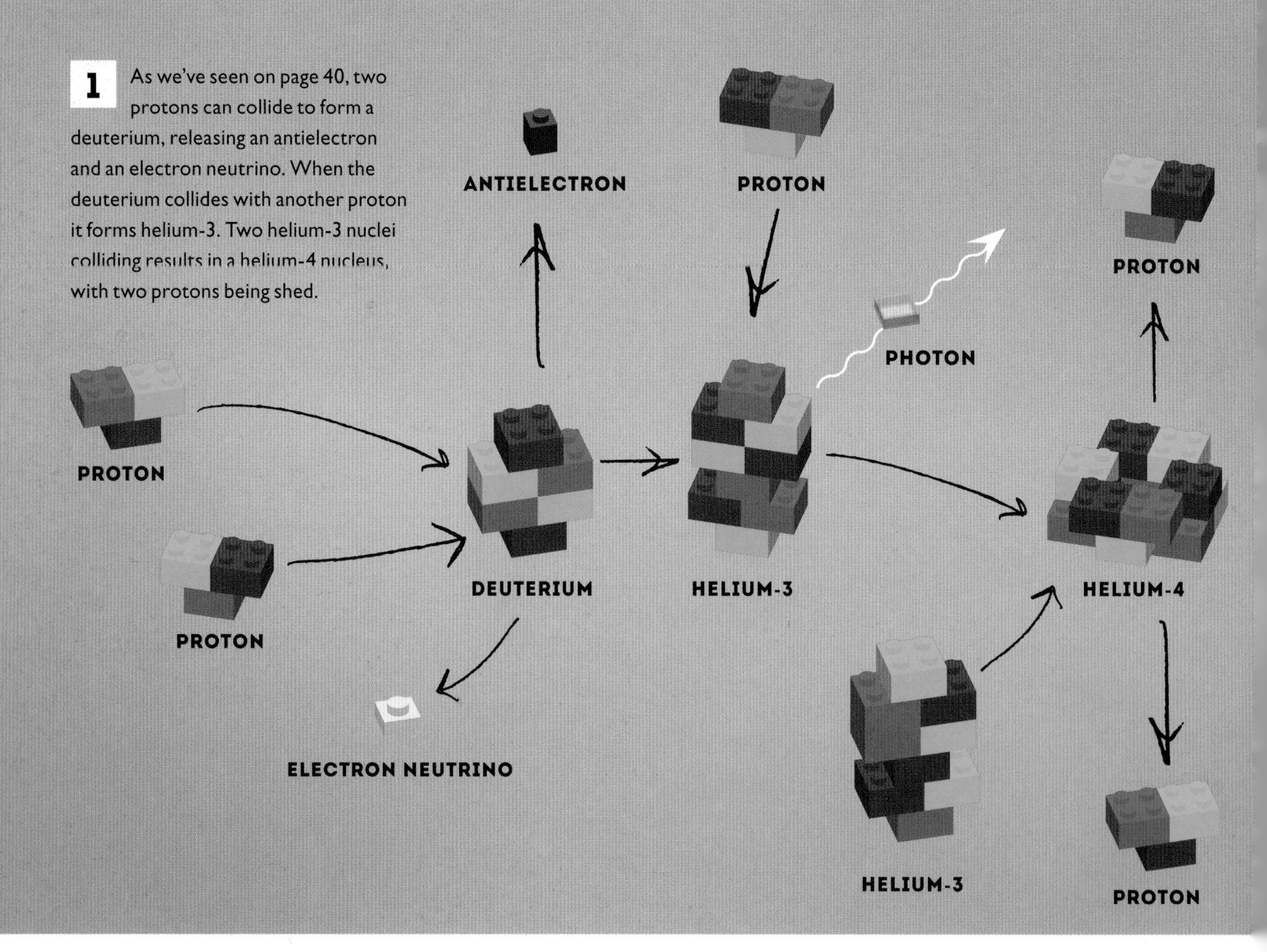

2 Once helium-4 is produced in a star it can collide with yet another helium-3 and the two nuclei fuse to form a beryllium-7 nucleus (four protons and three neutrons).

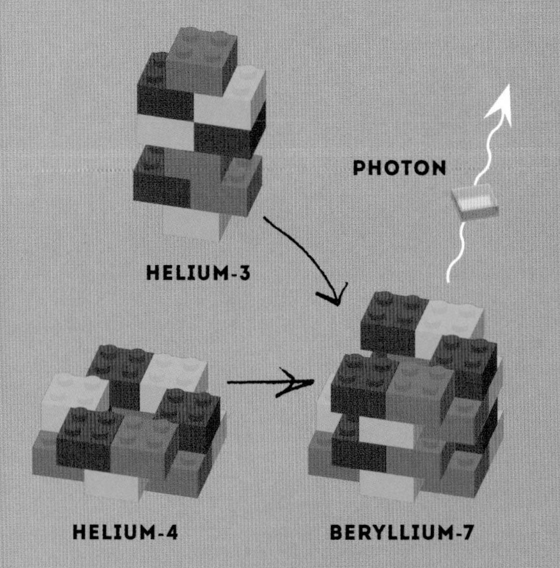

3 Beryllium-7 is unstable and soon captures a nearby electron to convert a proton to a neutron, forming a stable lithium-7 nucleus (three protons and four neutrons). An electron neutrino particle is emitted in this process.

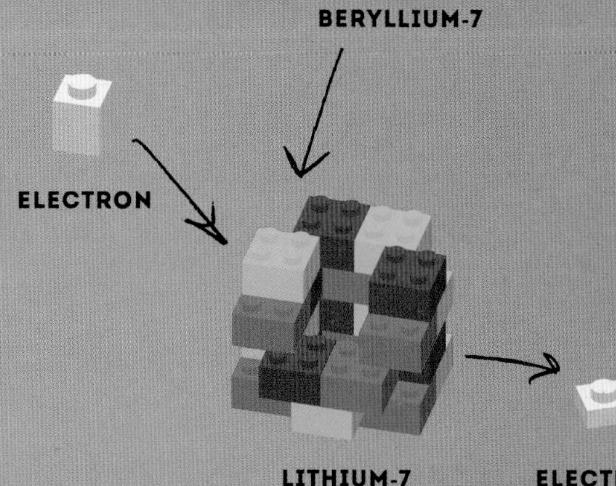

4 A proton colliding with lithium-7 forms, for a brief moment, beryllium-8; four protons and four neutrons. Beryllium-8 is unstable and soon disintegrates into two helium-4 nuclei.

Energy is released at each stage of this process in the form of photons of light. Although the end result of this chain is still helium-4, it demonstrates that stars are able to fuse protons together in different ways, creating the nuclei of new chemical elements. It is one of two main routes through which stars turn hydrogen into helium. The second is the topic of the next two pages.

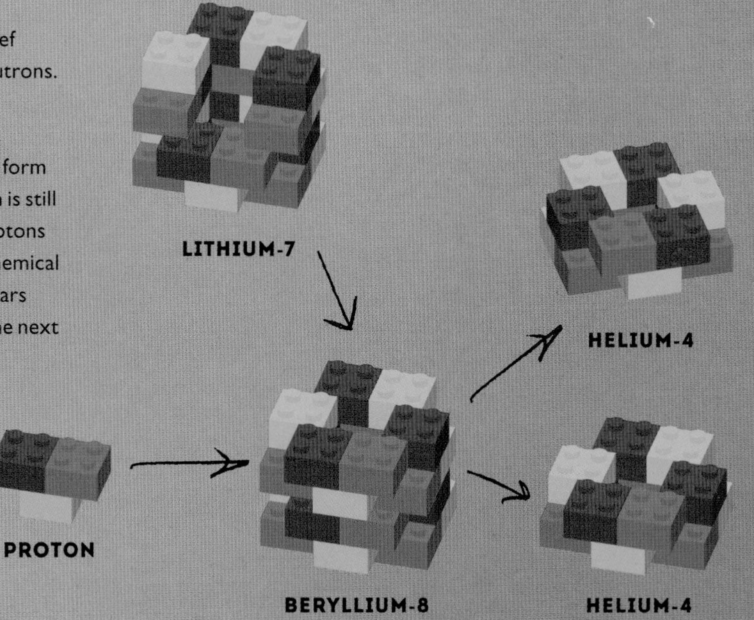

Leftover carbon from an older generation of star allows some stars to form helium in a different way

A NEW STAR IS BORN – THE CNO CYCLE

As time passes, stars come and stars go. When a star dies its remnants are expelled into space and may become the seed of a new star. These further generations of star will not be like the previous one, made entirely of hydrogen and some helium. They will also contain some heavier nuclei of other chemical elements: such as carbon (C), nitrogen (N) and oxygen (O). Our Sun contains quite a lot of these heavier chemical elements from another star. So, it is thought that one or possibly two generations of stars may have lived their lives in the same region of space that our solar system now occupies, and their remains contributed to the creation of the Sun and planets.

1 After colliding, a proton can fuse to a carbon-12 nucleus to form an unstable nitrogen-13 nucleus (7 protons and 6 neutrons).

2 Thanks to the weak nuclear force one proton then transforms into a neutron, ejecting an antielectron and an electron neutrino in the process, to form a stable carbon-13 nucleus (6 protons and 7 neutrons).

3 Add a proton to carbon-13 and you can create stable nitrogen-14 (7 protons and 7 neutrons), the same isotope of nitrogen that makes up 78% of the air we breathe on Earth.

EARTH'S NITROGEN

Nitrogen on Earth is actually a mix of 99.6% nitrogen-14 and 0.4% nitrogen-15.

6 A collision with yet another proton causes the nucleus to disintegrate, forming one helium-4 nucleus and a carbon-12 nucleus. This brings us back to the beginning of the cycle, ready for it to spin once more.

METALS IN STARS

If you ever hear an astrophysicist talking about metals in stars, they are not talking about the shiny elements which make up the bulk of the periodic table. Instead, astronomers call all elements heavier than helium a metal, even gases like nitrogen!

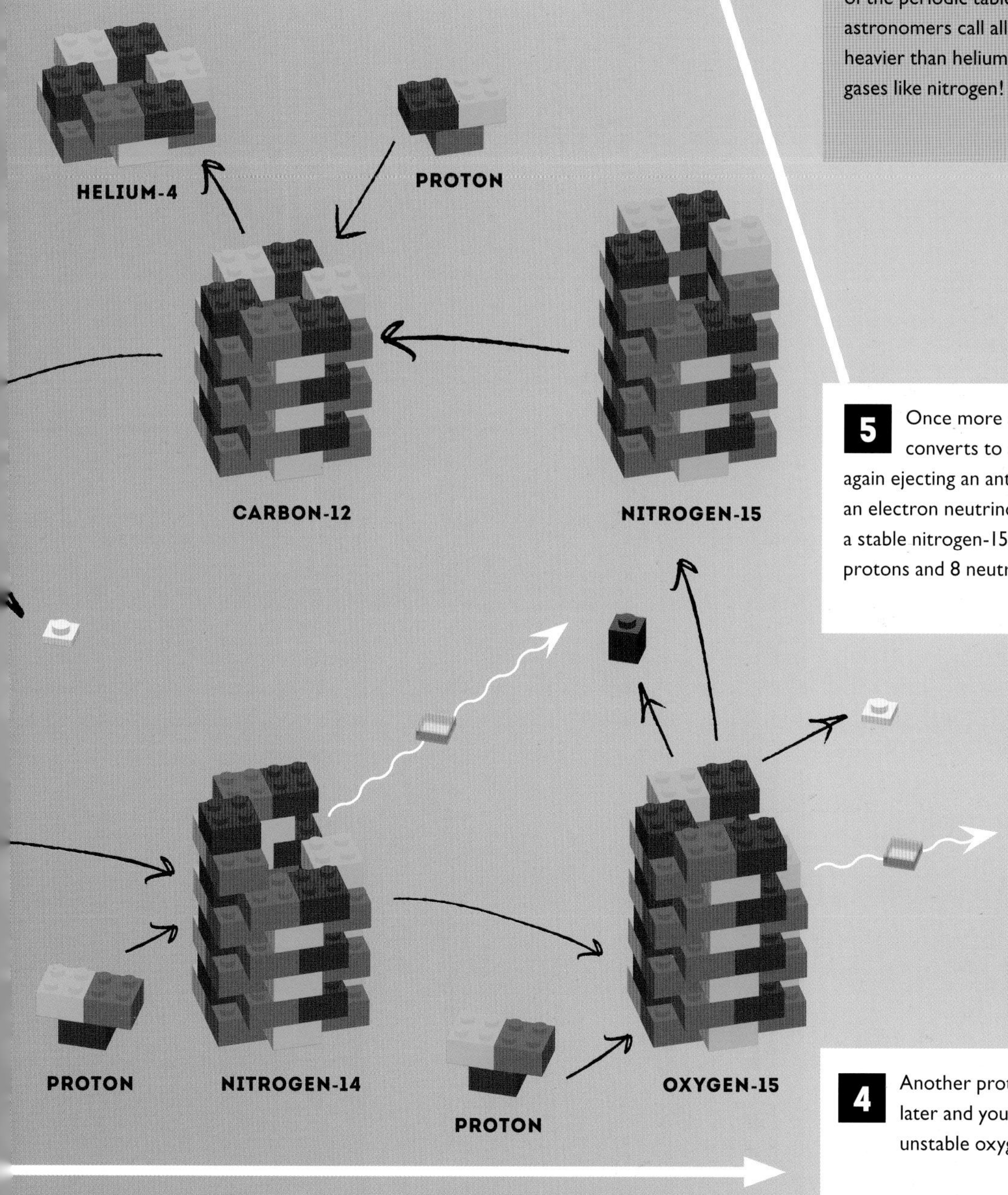

HELIUM-4

PROTON

CARBON-12

NITROGEN-15

5 Once more a proton converts to a neutron, again ejecting an antielectron and an electron neutrino. This forms a stable nitrogen-15 nucleus (7 protons and 8 neutrons).

PROTON

NITROGEN-14

PROTON

OXYGEN-15

4 Another proton collision later and you create unstable oxygen-15.

When there are no longer enough protons to continue fusion, the star collapses inward, fusing the helium into heavier elements

TRIPLE ALPHA PROCESS

The proton–proton chain and the CNO cycle fusion processes reduce the number of protons in the core of the star while increasing the amount of helium-4. There comes a point in a star's life where there are not enough protons in its core colliding in the short time needed to continue fusion. At this point all fusion reactions in the core of a star stop and the core becomes *inert*, another word for unreactive.

When this proton to helium-4 fusion stops in the core, gravity takes hold of the star and the gas from which the star is made begins to collapse inward. This increases the temperature and pressure within the star, leading to more energetic and frequent collisions between particles. This increase in temperature is enough to ignite proton–proton fusion in areas outside the core, forming a layer around the inert helium-rich core.

During this time the core continues to heat up. With a core temperature of 100 million degrees, positive electrically charged helium-4 nuclei can overcome their repulsion of each other and fuse together.

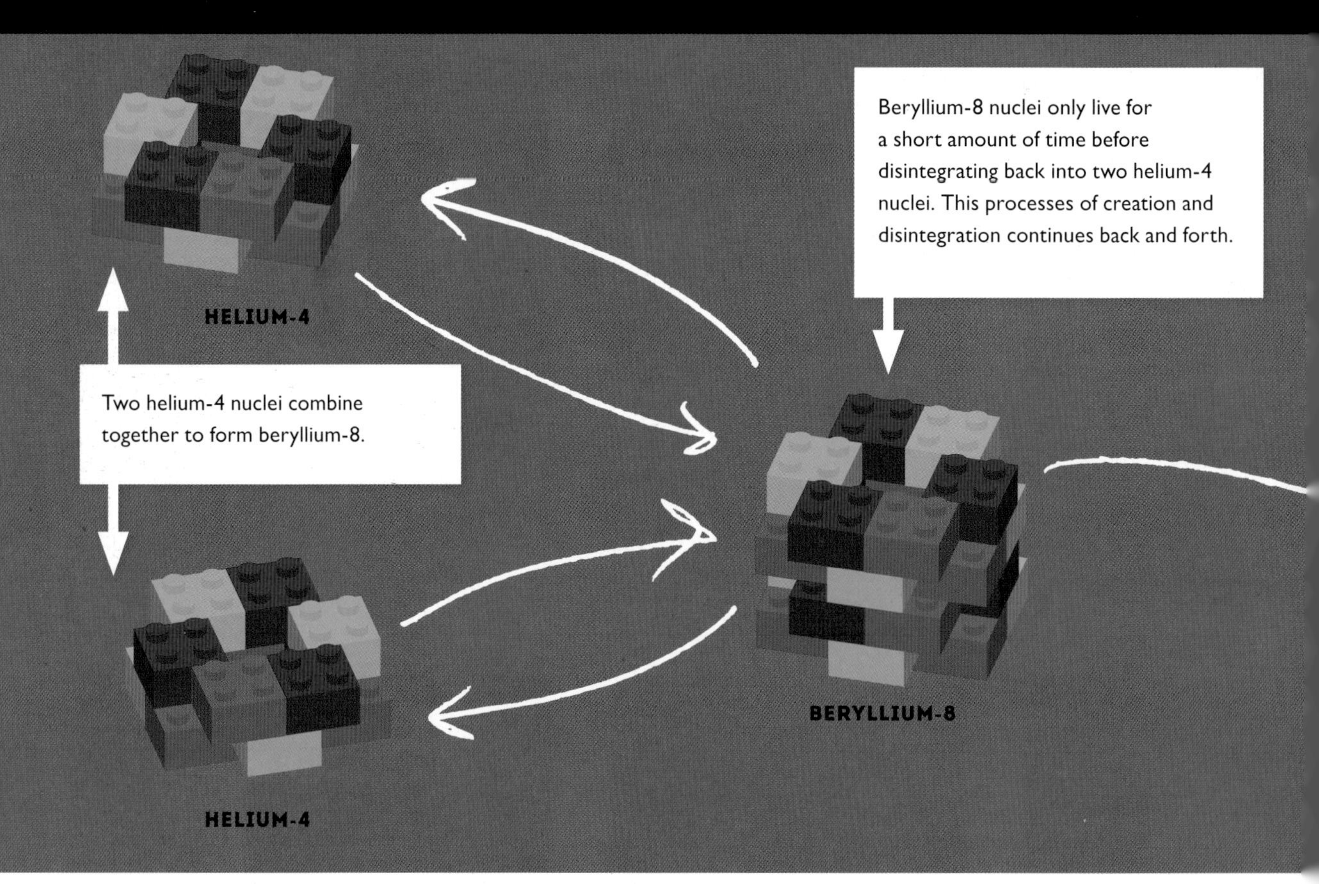

Although short lived, increased pressure in the star's core allows a third helium-4 nucleus the opportunity to collide with the fleeting beryllium-8 nucleus. Add a helium-4 nucleus to a beryllium-8 nucleus and you get a stable carbon-12 nucleus.

The Pillars of Creation are giant columns of hydrogen gas and dust which contract to form stars.

HELIUM-4

CARBON-12

ALPHA PROCESS

Wondering why is it called the alpha process? It's to do with the history of radiation and another name for helium-4. See pages 86–7 for the answer.

Helium nuclei continue to fuse to carbon nuclei to build up different heavy chemical elements

ALPHA LADDER

For stars with a mass less than eight times that of our Sun carbon-12 is the end of the fusion line. Temperatures at the core of such a star never get high enough for helium-4 nuclei to overcome the large electromagnetic repulsion of carbon-12 with its plus six positive electric charge. Stars with mass eight times the Sun or greater, however, continue to fuse. A continued increase in temperature and density means that helium-4 can continue to collide and combine with the heavier nuclei via the alpha process.

Depending on the size of the star this process can continue to create larger nuclei. But as the nuclei become larger, and therefore contain more protons and positive electric charges, it is ever-more difficult to get them to collide with helium-4 nuclei. For this reason, it is far more difficult to produce elements using this method which are much heavier than neon-20. As each of these stages of fusion occurs at a very low rate it means that they do not contribute much to the overall energy produced by the star – most of it continues to come from proton–proton and helium fusion.

ODD AND EVEN

Because most of the fusion to create heavier elements involves helium-4, nuclei of elements which have an even number of protons in them are more common than elements with an odd number. Odd-number elements require the weak nuclear force and a conversion between proton and neutron.

1 Add a helium-4 nucleus to a carbon-12 nucleus and you create oxygen-16, the most common isotope of oxygen found in the Universe.

CARBON-12

HELIUM-4

PHOTON (LIGHT)

OXYGEN-16

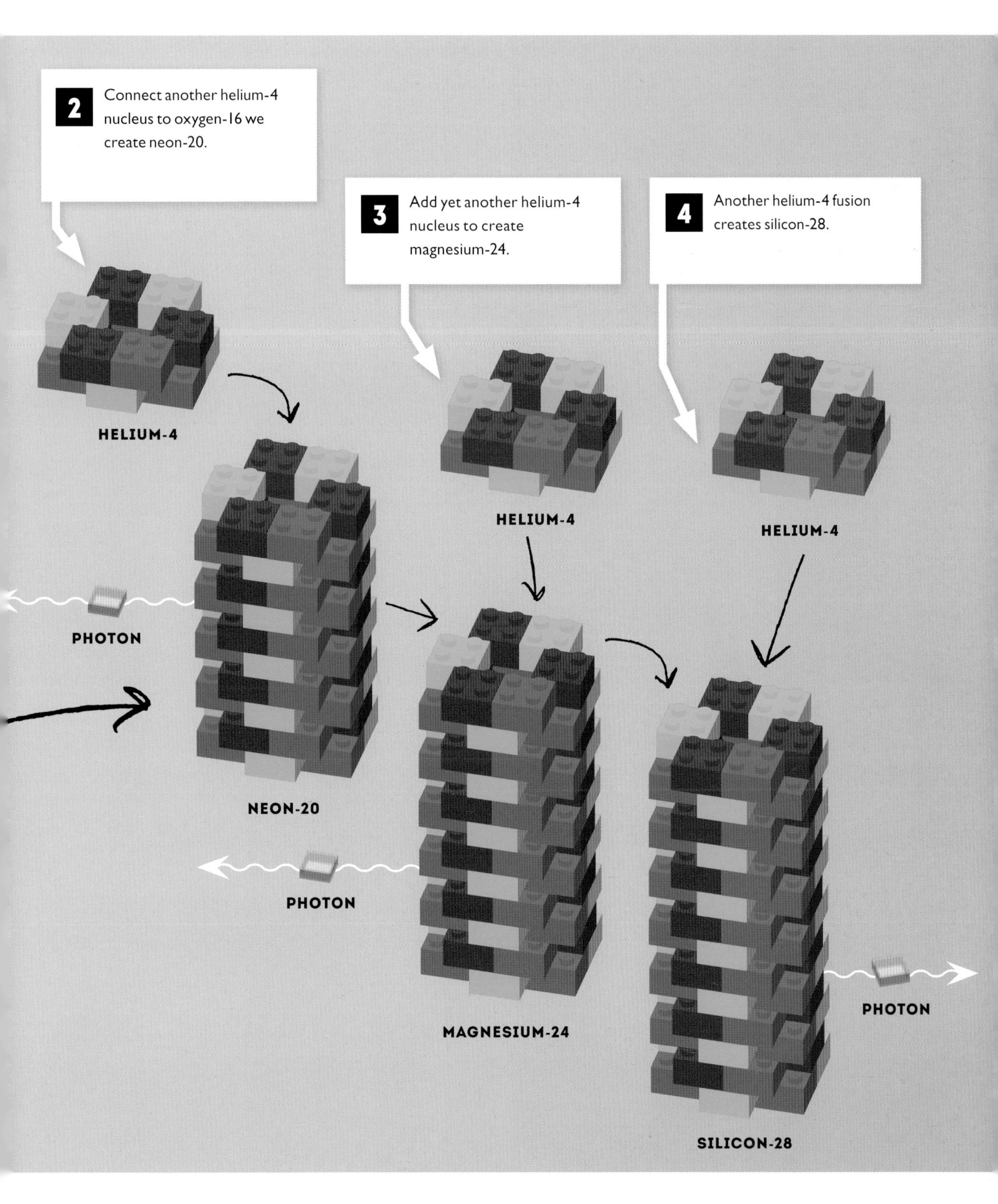
2
Connect another helium-4 nucleus to oxygen-16 we create neon-20.
3
Add yet another helium-4 nucleus to create magnesium-24.
4
Another helium-4 fusion creates silicon-28.
HELIUM-4
HELIUM-4
HELIUM-4
PHOTON
NEON-20
PHOTON
MAGNESIUM-24
PHOTON
SILICON-28

Soon most of the helium in the core is used up, and carbon nuclei become the next fuel for fusion

CARBON BURNING

Eventually helium burning, as all good things, comes to an end in the core of a supermassive star. This happens in much the same way as the end of other fusion types, due to a reduced concentration of things to fuse. The inert carbon-rich core heats up as the star once more begins to collapse. Helium-4 fusion now ignites outside the core, joining the outer proton–proton layer. The star is like an onion with layers of different fusion processes all going on. The central core of a star, with mass about eight times that of our Sun or greater, heats up to over 500 million degrees Kelvin. At this temperature carbon-12 nuclei have enough energy to overcome the repulsion they feel for each other because of their six positive electrically charged protons. When they meet they fuse to create heavier nuclei.

CARBON-12

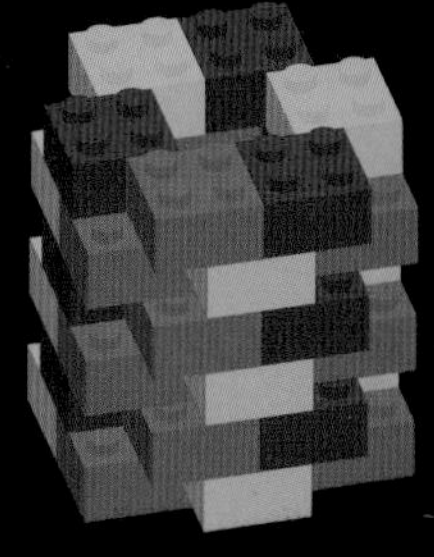

CARBON-12

NEON-20

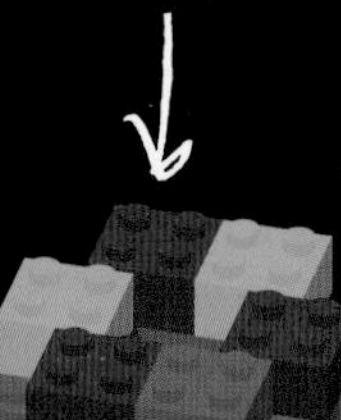

HELIUM-4

NEON-20
The most likely reaction is carbon-12 nuclei coming together to form a neon-20 nucleus and a helium-4 nucleus.

CARBON-12

CARBON-12

MAGNESIUM-24

MAGNESIUM-24

Simply connecting two carbon-12 nuclei together to form a magnesium-24 nucleus is one route of carbon fusion but it is far from the most likely. The magnesium-24 that would be created would be highly energetic and must lose a large amount of energy to survive. The most effective way of losing energy is by releasing particles, which is the reason that nuclei other than magnesium-24 are more likely to be created.

LIKELIHOOD OF REACTIONS

You will read later in the book that despite some reactions between particles being possible, the strength of the different force regularly dictates the likelihood that they will occur.

As the carbon nuclei run out, gravity collapses the star further, heating the core and fusing oxygen nuclei

OXYGEN BURNING

High temperatures inside supermassive stars not only allow for nuclei to fuse but also produce photons energetic enough to split nuclei apart. The high proportion of oxygen that is found in the cosmos, compared to that expected from just fusion, is due in part to the splitting of neon-20 nuclei by photons to produce oxygen-16 and helium-4. This occurs just before a star again begins to contract and heat up once more, as the rate of carbon burning decreases. The core rises to a temperature around 2 billion degrees Kelvin and now the increased number of oxygen-16 nuclei have enough energy to get close enough together to fuse. A number of different end products can be formed but we'll look at the two most likely here.

SILICON-30

As with carbon burning, the naively most obvious end product would be sulphur-32, with the same number of protons and neutrons as two oxygen-16 nuclei combined. However, the resulting sulphur-32 would be too high in energy and so protons, neutrons, and nuclei have to be ejected, forming silicon-30.

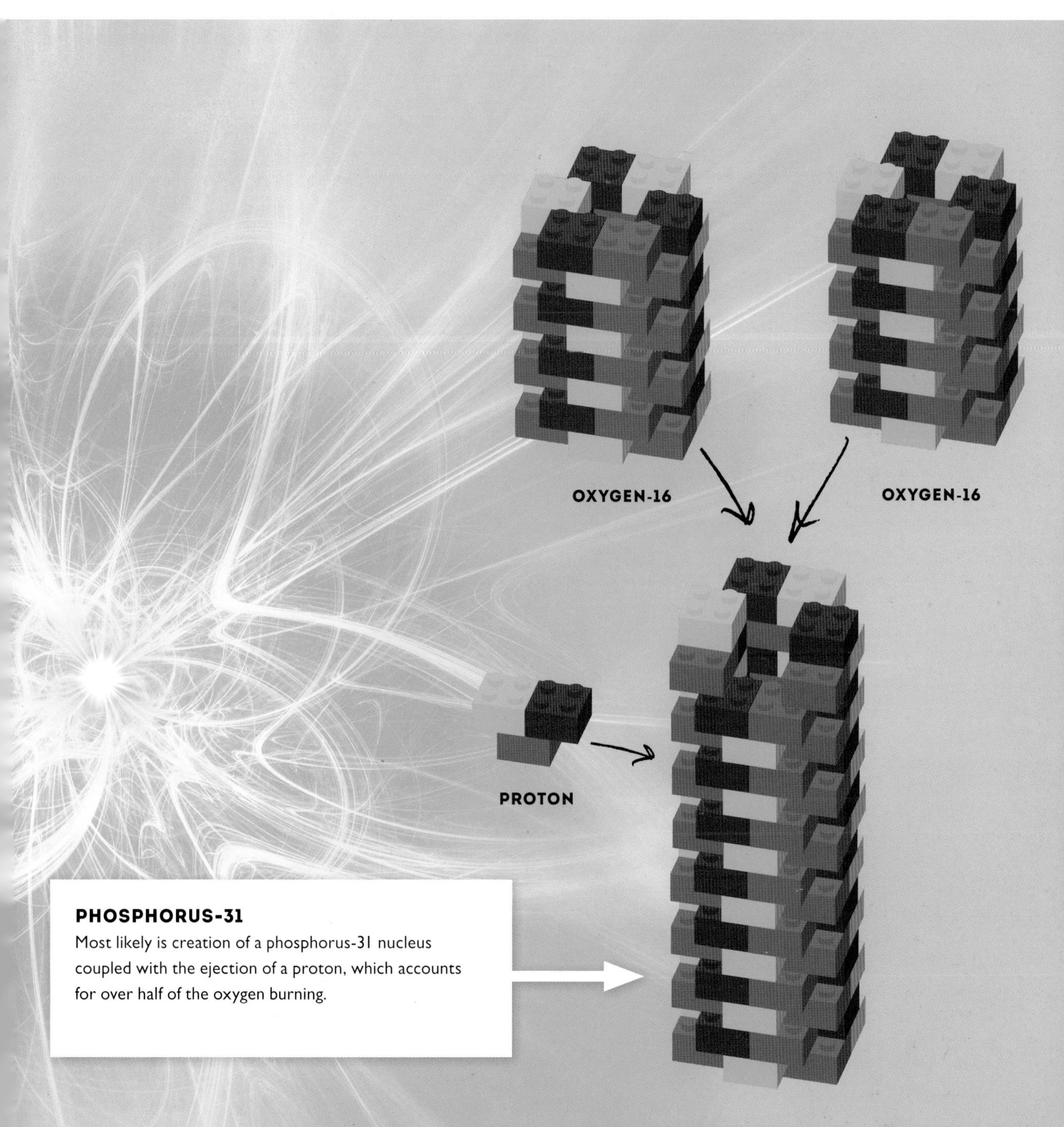

PHOSPHORUS-31

Most likely is creation of a phosphorus-31 nucleus coupled with the ejection of a proton, which accounts for over half of the oxygen burning.

Energetic photons break nuclei into helium-4, providing fuel for further fusion

SILICON BURNING

After oxygen burning winds down, the core of a star is rich in silicon-28 nuclei. As it contracts and continues to heat it reaches the right temperature for silicon burning to occur. While carbon and oxygen burning involved the direct fusion of the nuclei, silicon burning happens in a slightly different way. Although it is possible, two silicon-28 nuclei do not have the right conditions to fuse directly to form nickel-56. Instead, at this temperature photon particles of light have a high enough energy to 'melt' the silicon-28 nuclei. At each stage of this melting a photon chips away at the nucleus, knocking off a helium-4 nuclei each time.

In stars with a mass greater than 11 times that of the Sun helium-4 nuclei can continue to fuse to nuclei, increasing their size. This can continue up to nickel-56 via argon-36, calcium-40, titanium-44, chromium-48 and iron-52, adding two protons and two neutrons each time.

Silicon is 'melted', breaking it down into helium-4 nuclei constituent parts. These helium-4 nuclei are then fused onto other silicon-28 nuclei to build larger nuclei.This melting chain, the reverse of the fusion on pages 50–1, enriches the core once more with lots of helium-4 and allows the earlier alpha process to continue.

The new-found helium-4 adds two protons and two neutrons to a silicon-28 nucleus, first forming sulphur-32.

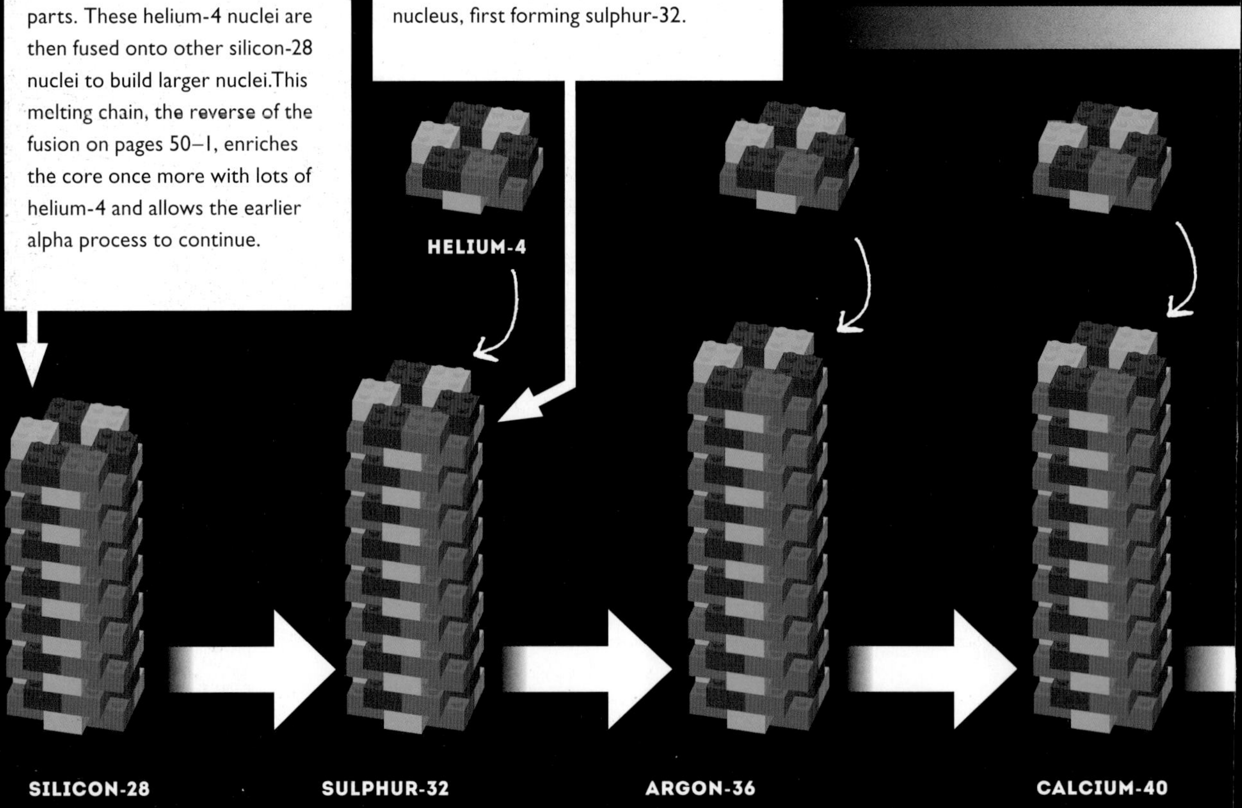

HELIUM-4

SILICON-28

SULPHUR-32

ARGON-36

CALCIUM-40

ALPHA PROCESS

The alpha process in stellar fusion is the repeat addition of a helium nucleus (alpha particle) to build heavier nuclei.

Each stage is a two-way reaction, with some nuclei getting heavier while some melted smaller by a photon. The fusion of helium-4 with heavy nuclei ends at nickel-56 because creating a nucleus with more than 56 protons plus neutrons would require and not release energy.

Why 56 protons plus neutrons? It is all to do with the strength of the electromagnetic and strong forces. With this number of protons and neutrons the strong force binding the nucleus together overpowers the electromagnetic force more than in any other nucleus, making it the most stable possible.

TITANIUM-44 **CHROMIUM-48** **IRON-52** **NICKEL-56**

The s-process and r-process build elements heavier than nickel-56 through neutron capture and their decay to protons

THE HEAVY ELEMENTS

Creating the many elements heavier than iron and nickel requires something different to fusion of nuclei, the capture of neutrons. After a nucleus absorbs a handful of neutrons some decay to form protons, emitting an electron and antineutrino in the process, forming a new chemical element. There has to be a high density of neutrons surrounding any nuclei if they are to be captured.

Stars with mass ten times that of the sun and higher have high enough concentrations of neutrons at their core. Still, a tiny chance of capture occurring means that creation of new nuclei in this way is a slow process, which scientists imaginatively abbreviate to the s-process.

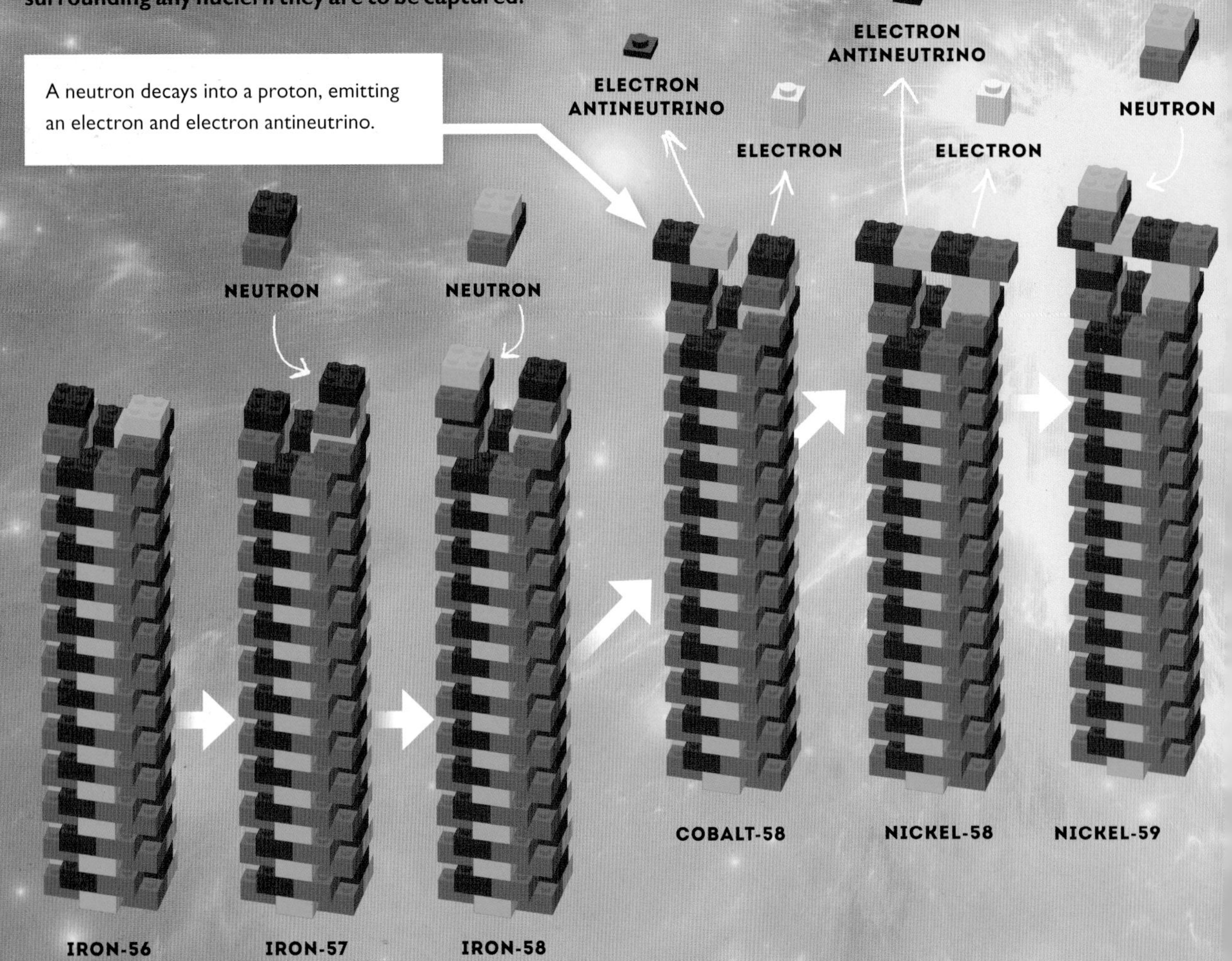

It is possible to build nuclei up in size all the way to create the heaviest stable nucleus, bismuth-209. As soon as a bismuth-209 nucleus is made, any further neutron captures pushes the process around in a cycle that always ends up with bismuth-209. To add more neutrons to make nuclei larger requires a far higher density of neutrons than exists within a living star.

BISMUTH-209

LEAD-206

LEAD-207

LEAD-208

The absorption of two or more neutron simultaneously is the only way of overcoming this terminating cycle. This can only happen in the extremely neutron dense environment provided by the cataclysmic death of a supermassive star, an event known as a supernova. When fusion processes in a star end, gravity wins over, the dense cloud of nuclei, electrons, and photons rapidly collapses in on itself. The increase in pressure crams more and more particles into a tighter space, allowing nuclei to capture lots of neutrons. This death march happens over a course of just minutes and so this is known as the rapid process, or r-process for short. The unstable nuclei created at various stages of r-process neutron capture do not have time to disintegrate into lighter nuclei, allowing them to rapidly build up in size, like a snowball rolling down a snow-covered hill. They can get pumped up to nuclei that contain up to 270 protons and neutrons. These nuclei are of course very neutron rich and, after the density of neutrons decreases after the supernova, they rapidly rearrange, converting neutrons into protons through the weak nuclear force.

URANIUM

Uranium has the title of heaviest naturally occurring element, and most of it on Earth (99.3%) exists as uranium-238. It has earned this title because although the r-process of neutron capture may create heavier nuclei, all of these have lifetimes so short that they do not exist in measurable amounts for very long.

A symmetry is any group of transformations of an object which leaves it unchanged

SYMMETRIES

In this chapter we will discuss in detail the most familiar of atomic forces, the electromagnetic force. Before we describe the force we must first talk a little of the formulation of the Standard Model, how it explains particles and interactions, and how the strange quantum behaviour of particles leads to the force of electromagnetism.

In physics we observe that as an experiment evolves, certain properties that we measure remain unchanged. One is that the total energy of any group of particles (a closed system) is always conserved. It is the same before and after an event. You may have heard this put in another way; energy cannot be created or destroyed, only transferred. Another conserved measurable of the group of particles is their momentum, which is (at low velocities at least) a combination of velocity and mass. In the real world of course, it is difficult to truly isolate one group of particles from another. Despite this there has been no observation of energy or momentum not being conserved.

Emmy Noether was a fantastic mathematician who died shortly arriving in the USA after fleeing Nazi-occupied Germany. Her legacy though is immortal. She proved that every conserved physical quantity that exists in Nature (like energy, momentum and electric charge) is the consequence of some symmetry in our Universe.

MASS

Einstein's special relativity flies in the face of another high-school 'conserved' quantity – mass. Mass is only conserved when things are travelling well below the speed of light. Particles travel at high speed, however, and so it is energy that is ultimately conserved and mass is just another form of energy.

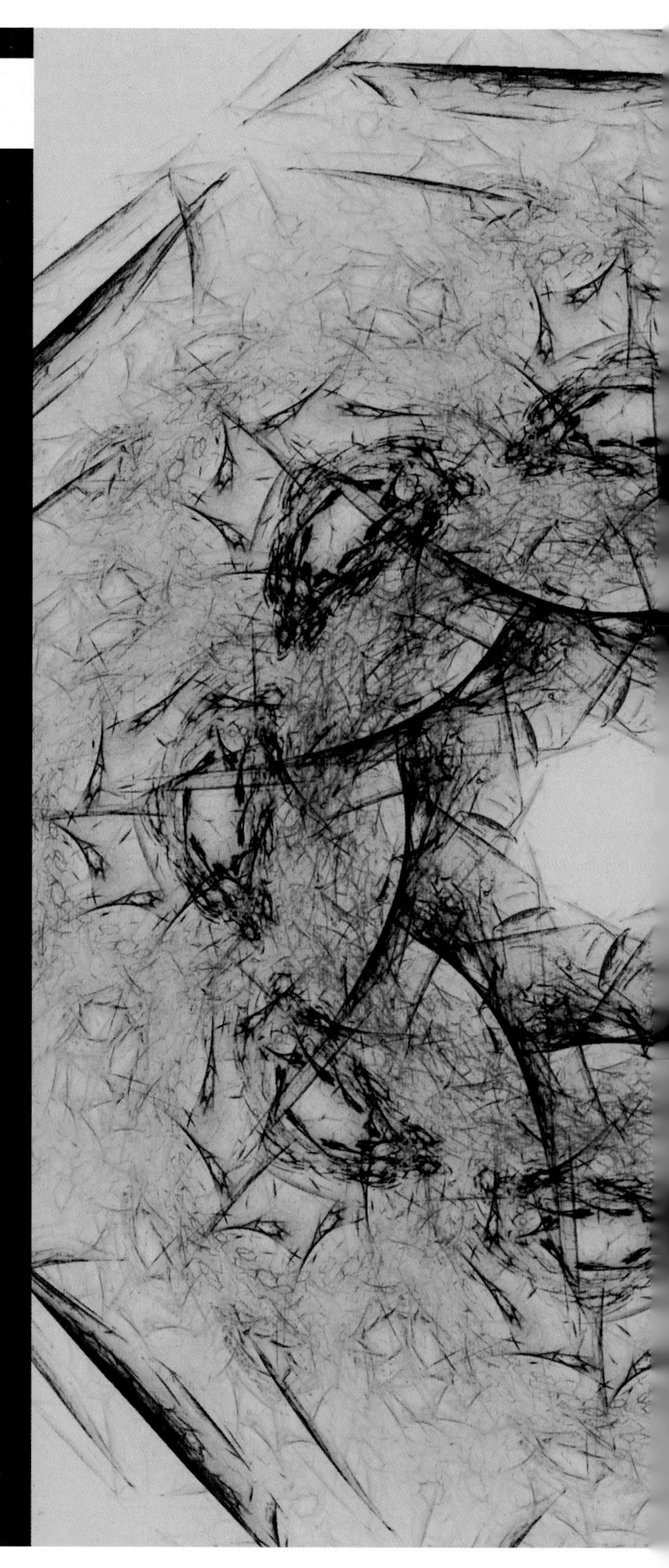

A 2X2 PLASTIC BRICK SHOWS A NUMBER OF DIFFERENT SYMMETRIES

Rotational – turn the block by 90°, 180°, 270° or 360° and you find yourself looking at an identical block.

It also has planes (imaginary slices) of mirror symmetry – reflect one half of the block through either centre line and you get the same block.

The block is also symmetric if reflected in a mirror – it has reflection symmetry.

We might also find that while a single transformation by itself is not a symmetry, some combination of a number of transformations does leave an object exactly the same. These combinations form a symmetry and, as Emmy Noether tells us, represent a conserved quantity.

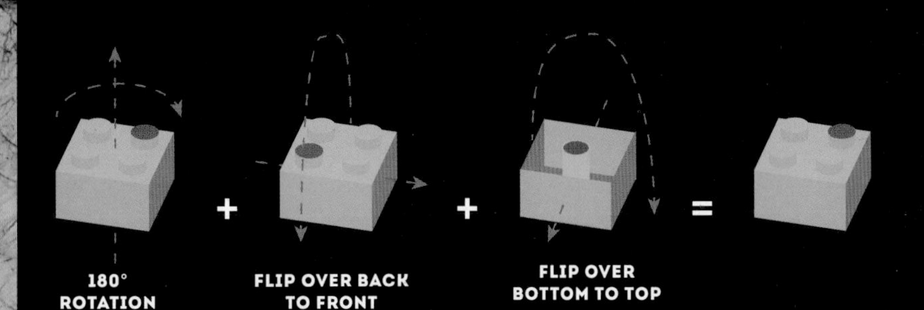

Energy is conserved because the laws of physics are symmetric in time – the same piece of physics holds today as it did yesterday. This suggests that physics is legitimate whether played forward or backward in time. Momentum is conserved because of a symmetry of space – it does not matter in which direction you calculate physics because all space looks the same. This suggests that an experiment done in London will get the same results as the exact same experiment done in New York.

These symmetries ultimately led Albert Einstein to his most famous equation equating energy and mass, with the exchange rate being the fastest possible velocity – the speed of light – squared (see pages 32–3).

Electric charge defines how particles interact with each other through the electromagnetic force

ELECTRIC CHARGE

Many symmetries similar to space and time exist in the mathematics of the Standard Model, each with a conserved quantity. Electric charge is conserved in particle interactions and is a result of a symmetry, one which represents the electromagnetic force. Benjamin Franklin identified electric charge as either positive or negative in the 18th century.

Electric charge is an internal property of a particle – unaffected by changes in space or time. Electromagnetism arises from a one-dimensional symmetry, which means electric charges of any particle can be written on a one-dimensional number line. Some are negative, like the electron, while others are positive, like the proton, but all electric charges will lie somewhere on the line.

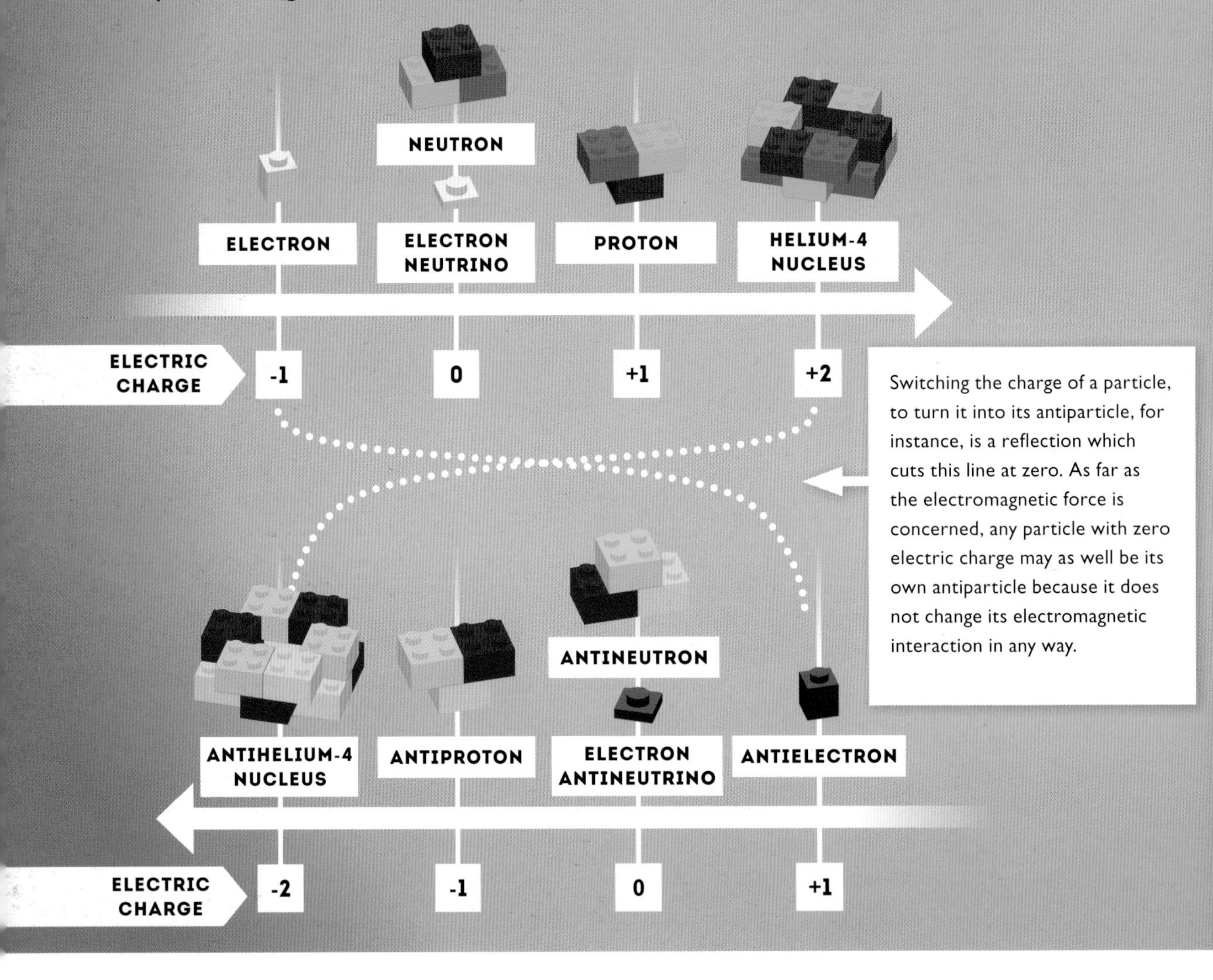

Switching the charge of a particle, to turn it into its antiparticle, for instance, is a reflection which cuts this line at zero. As far as the electromagnetic force is concerned, any particle with zero electric charge may as well be its own antiparticle because it does not change its electromagnetic interaction in any way.

Trying to separate two opposite electric charges is like dragging two sleds up snowy hills on opposite sides of a valley – the increase in height increases the potential energy between them. A sled has the potential to gain speed (kinetic energy) when falling under gravity. Electric charges have the potential to gain speed when attracted to or repelled from other electric charges. Attraction and repulsion just change the shape of the electromagnetic potential 'hills' that charges are pulled up.

Like charges always repel, as if they want to go down opposite sides of a mountain away from each other to lower potential energy.

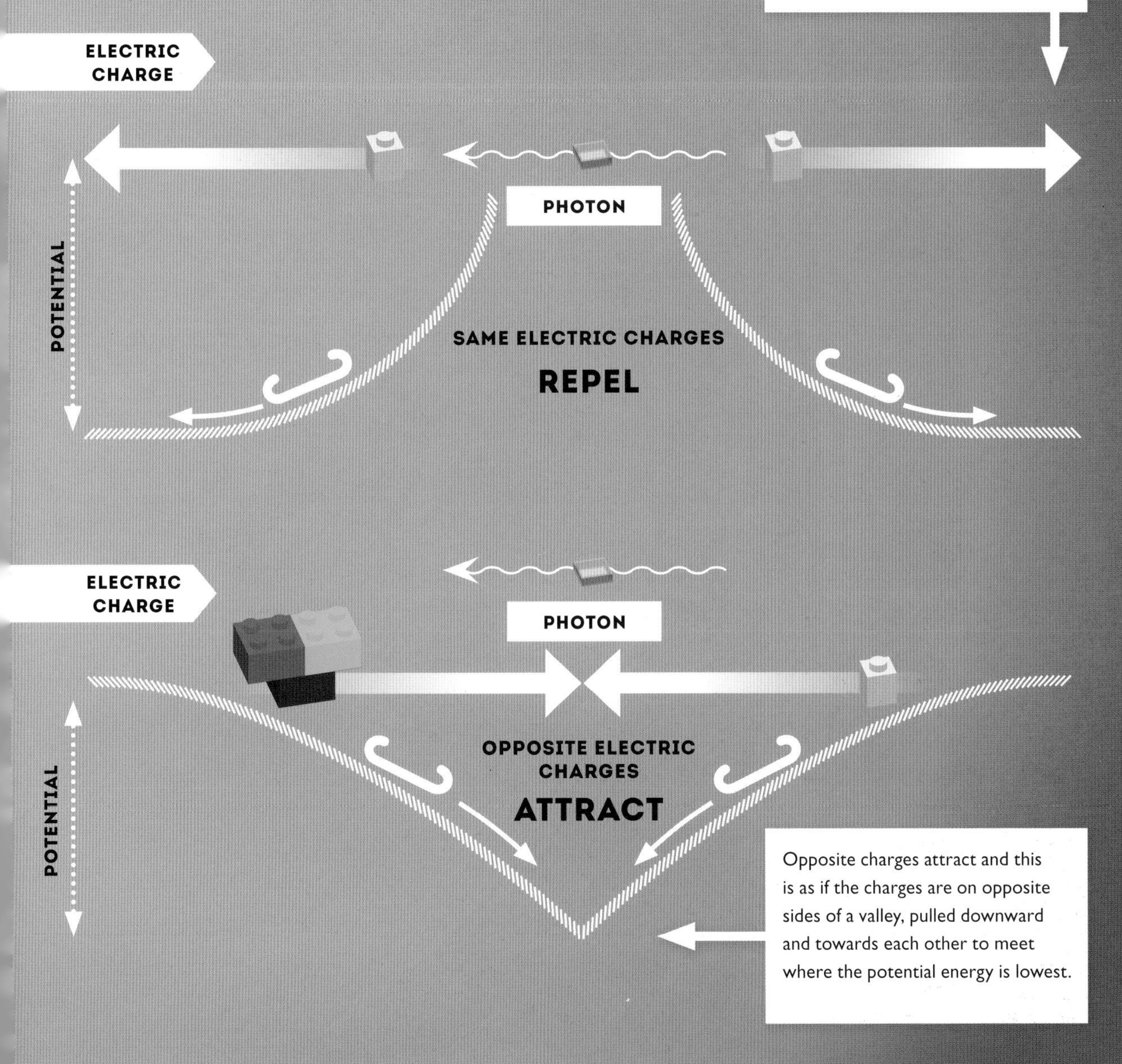

Opposite charges attract and this is as if the charges are on opposite sides of a valley, pulled downward and towards each other to meet where the potential energy is lowest.

Particle accelerators use electric fields to accelerate electrically charged particles to high speeds

PARTICLE ACCELERATORS

Electrical potential is measured in volts, V. I mentioned in chapter 1 that an electron accelerated by a potential of 1 V would gain a kinetic energy of 1 eV. It gains this energy by falling down the potential set-up much like a sled gaining kinetic energy down a hill. Make the hill bigger and the kinetic energy increases more. Accelerate an electron across a potential of some number of volts N and it will gain N eV of kinetic energy.

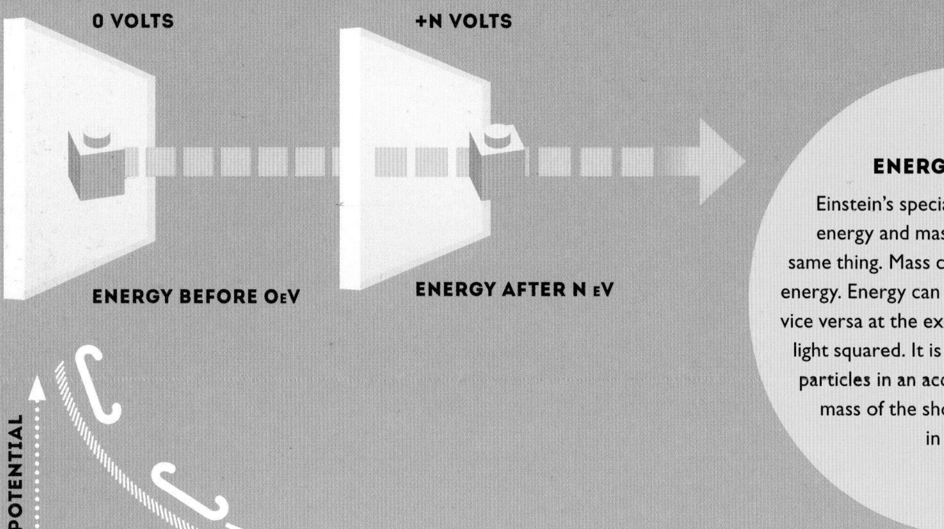

ENERGY AND MASS

Einstein's special relativity tells us that energy and mass are two forms of the same thing. Mass can be thought of as frozen energy. Energy can be exchanged for mass and vice versa at the exchange rate of the speed of light squared. It is the extra energy given to particles in an accelerator that creates the mass of the showers of new particles in collisions.

The Large Hadron Collider (LHC) accelerates protons to around 6.5 TeV. To do this in one go would require 6.5 million million volts. Such a large electric potential would tear the proton apart. Instead, to reach such high energies you have to move the hill. If the hill is constantly rising while you are falling then you would never reach the bottom, like a jogger never reaching the end of the treadmill. Particles on a moving electric potential therefore always experience falling. As they move faster so must the wave of electric potential. Otherwise the particle will find itself no longer falling; instead it will be either at the bottom of the hill or worse still on the upward slope slowing down.

The wavelength of the accelerating electric field has to get smaller and smaller to ensure it keeps up with pushing an ever-faster particle.

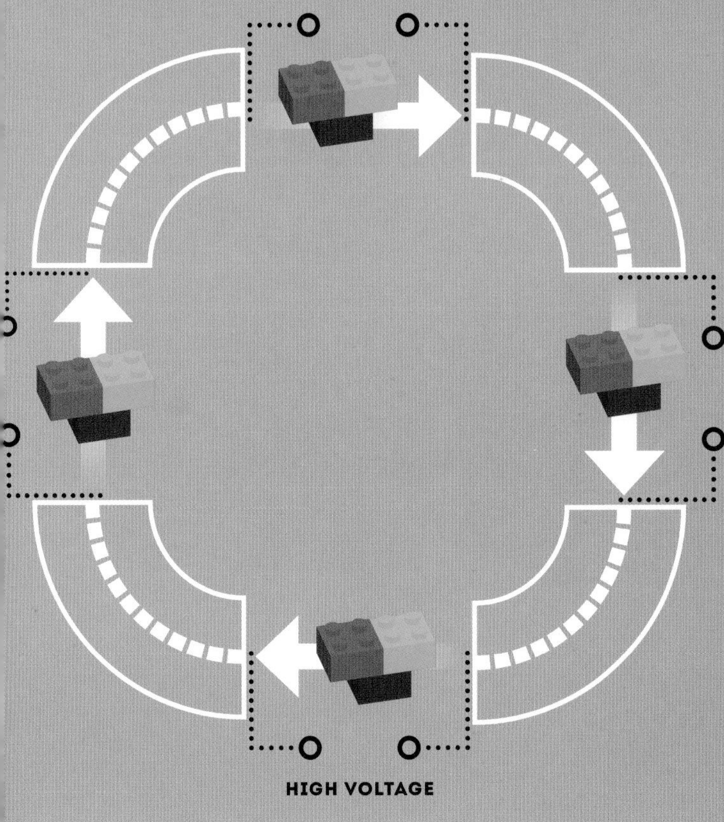

The Large Hadron Collider is a synchrotron particle accelerator 27 km around. It can accelerate protons to 6.5 TeV.

You can use this method to accelerate particles in a straight line in machines called linacs (abbreviation of linear accelerator) or around a ring.

In a linac, once the particles fly out the end they are gone. Use magnets to bend particles into a circle, though, and you can keep the wave going, like a Mexican wave rippling around a football stadium. As long as the wave speeds up in synch with the particles they will gain more energy. These machines, of which the LHC is one, are aptly called synchrotrons.

In synchrotron particle accelerators particles are curved by magnets to form a ring. In certain sections high voltages are used to produce an accelerating electric field which pushes particle energies and speeds higher.

TERA

T stands for tera, 10^{12}, a trillion or a billion billion.

Quantum objects behave like brick particles when interacting but there are fields of possibilities in between

QUANTUM

Light is not a wave. Nor is it a solid brick-like particle. Its wave-like behaviour arises because of the motion of photons (particles of light). This strange quantum behaviour is best seen if light is shone through two small slits. Firing photons one by one through the slits you would expect, like dropping bricks through two small gaps, for them all to pile up in just two locations on the screen below, each representing a slit. Instead the picture that is built up is one you would expect only if waves were passing through the two slits, interfering with each other as they went: areas with lots of photons and areas with very few.

This is what you might expect when dropping bricks through two slits.

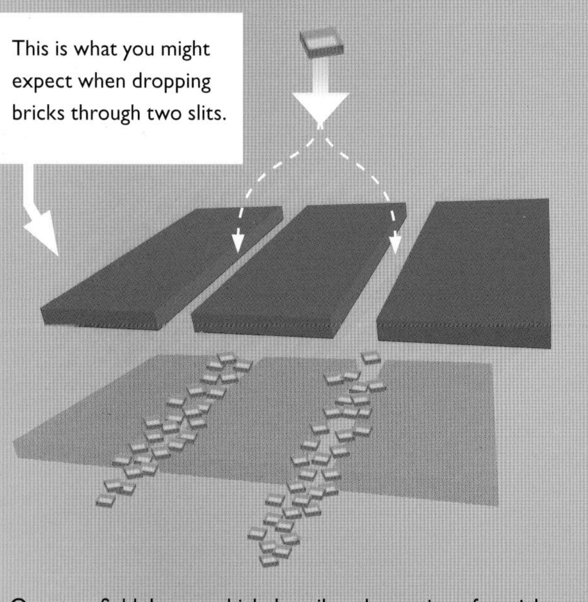

This is actually what happens, which can only happen if photons also behaved like a wave.

Quantum field theory, which describes the motion of particles and photons, explains this in the following way. When created, a particle (boson or fermion) is like a plastic brick with fixed properties; e.g. mass (size of brick). As it travels out into space, however, it behaves like something totally different, a field. The particle effectively spreads itself through all of space to try out all possible routes between two points, say A and B.

This can be imagined as the original particle splitting itself into an infinite number of different copies. Each copy tests a different route from point A and, when they finally meet at point B, they compare notes.

PARTICLE PEDOMETERS

But what do they compare? It turns out that particles have their own internal pedometers. Many people set pedometer targets to ensure they walk a certain distance each day. Rather than a fancy digital pedometer worn on the wrist, a particle's pedometer can be imagined like an analogue clock with the face ticking off steps.

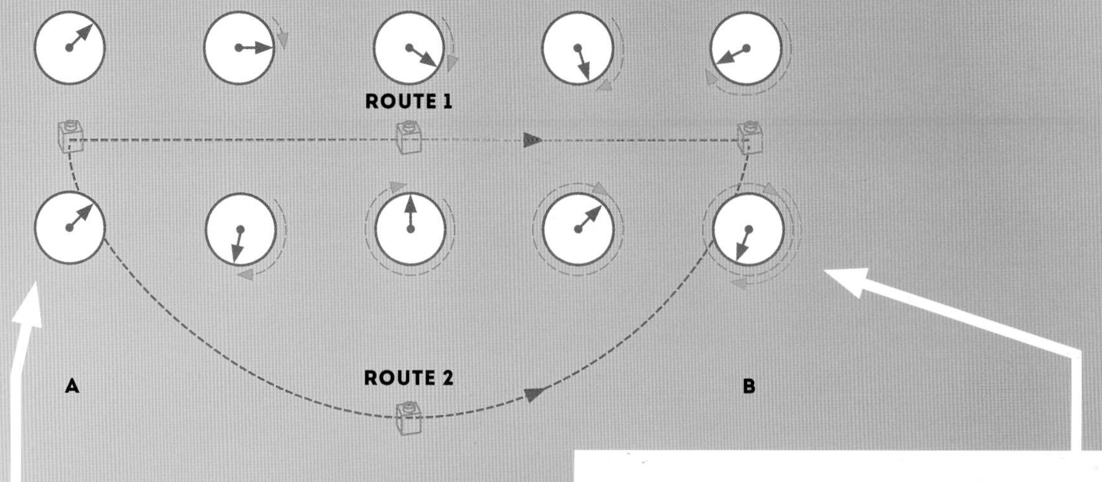

All copies of the particle are created at the same point and so begin life with synchronized pedometers (not necessarily zero).

If when travelling the path a particle reaches its target number of steps, the hand on the clock keeps spinning around, back past the original setting it had at point A. When two particles meet each other at point B there is no way of this particle proving that it has surpassed its target. The only thing which can be done is compare how different the readings are. This is known as the phase.

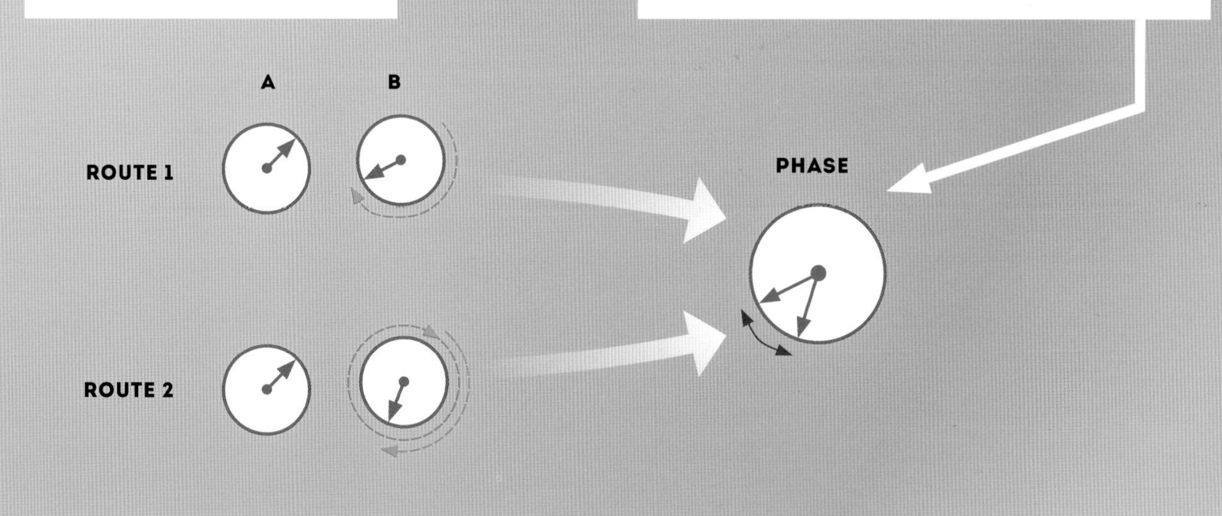

The comparison of the internal pedometers of two particles determines the probability that they interact with each other

INTERACTING QUANTUM FIELDS

Summing the phases of all routes a particle might take determines the probability the particle will travel that path between A and B. The smaller the difference in phase between routes (the more in phase), the higher the likelihood that path will be taken by the particle. Opposite phases cancel each other out entirely with phases in between, diminishing the overall likelihood. Once every possible path from A has been considered (C, D, E, F,...,Z etc.) we have the field, a whole spectrum probabilities of where on the screen below the particle is more and less likely to be seen. But, it is only when we measure it that we know which path it has taken.

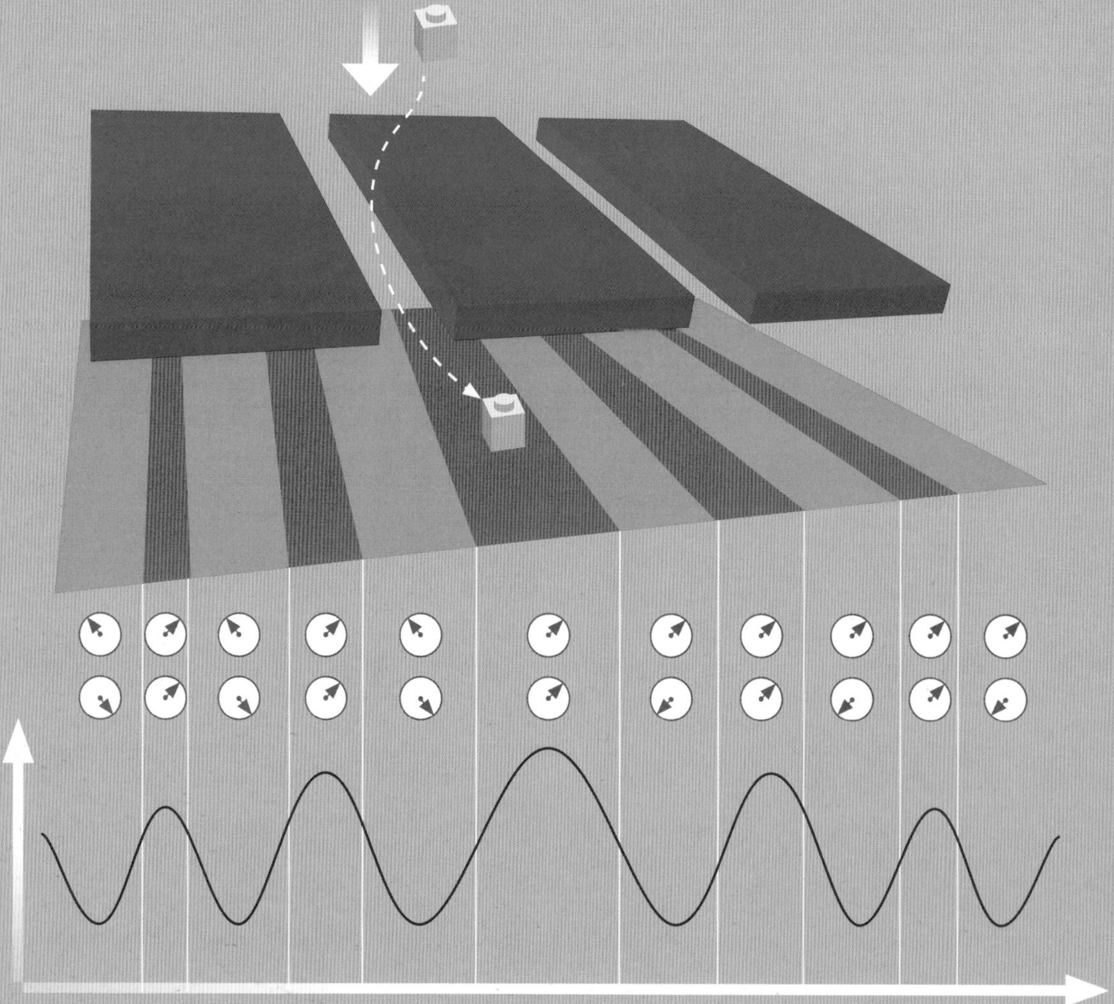

It's only ever possible to measure the probability that a particle will take a certain path, the phase difference between two pedometer readings, never the reading on a particle's pedometer directly – like using a stopwatch to measure the passage of time but never the time of day. A type of particle will behave the same no matter what initial reading it had on its pedometer – it is symmetric in the initial pedometer reading.

When two different particles meet they compare their pedometers. The probability that they will interact is not based on the initial or final pedometer reading, but instead is based on the change in readings after the particles have travelled along a path to that point. If two particles' readings have changed by a similar degree, they are more likely to interact.

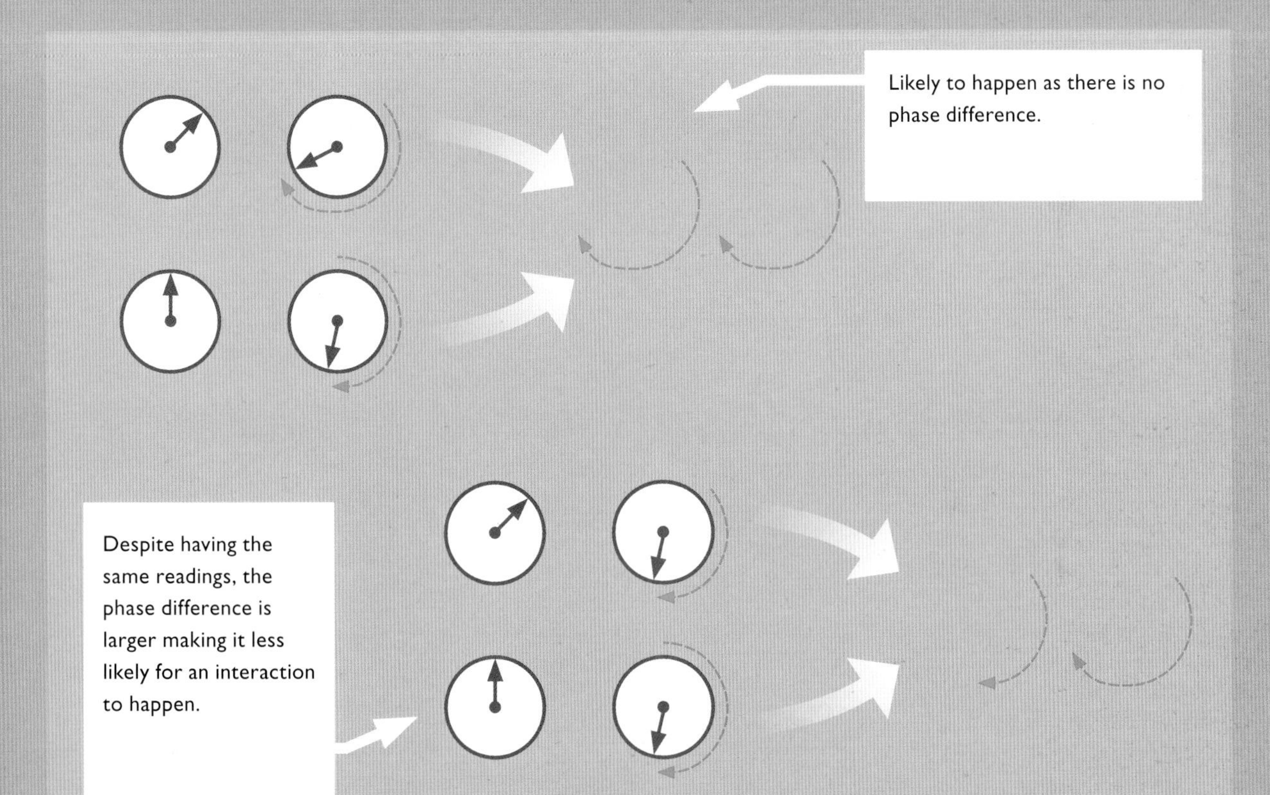

This phase difference comparison requires the particles to communicate – they need to know each other's original pedometer reading in order to work out by how much each other's reading has changed. This determines the phase difference and the probability that they will interact.

The original pedometers' readings are communicated through the electromagnetic field and if an interaction occurs then a photon is exchanged. The symmetry between these internal clock pedometers is the symmetry associated with the electromagnetic force.

Feynman diagrams describe the interactions between particles

FEYNMAN DIAGRAMS

Ask a particle physicist to explain particle physics and it won't be long before you see a Feynman diagram. These drawings are not just cartoons but a technical way of drawing interactions between particles – they are powerful mathematical tools. They were named after their creator, the enigmatic, Nobel Prize-winning, bongo-playing, US physicist Richard Feynman. Each diagram drafts possible paths and interactions that particles may take and is made from simple ingredients:

The diagram below shows an incoming electron emitting a photon in an electromagnetic interaction and travelling off in a different direction in space. All fermions with electric charge can interact with a photon in the same way – neutrinos cannot because with zero electric charge they cannot interact with a photon which carries the electromagnetic force.

Straight lines represent fermions with an arrow to show their direction.

Plastic brick to show the type of fermion (symbols are used in standard Feynman diagrams).

The vertical axis represents the position of particles in space.

The vertex is the point at which these lines meet and exchanges occur.

Wiggly lines represent the force-carrying boson (a photon in this example).

The horizontal direction represents the passing of time, from earlier times on the left to later times on the right.

The probability of an electron emitting or absorbing a photon at any vertex is a number called the coupling constant. For the electromagnetic force this is 1 in 137. The coupling constant defines the strength of the force at work and the value cannot be predicted, only measured directly from experiment.

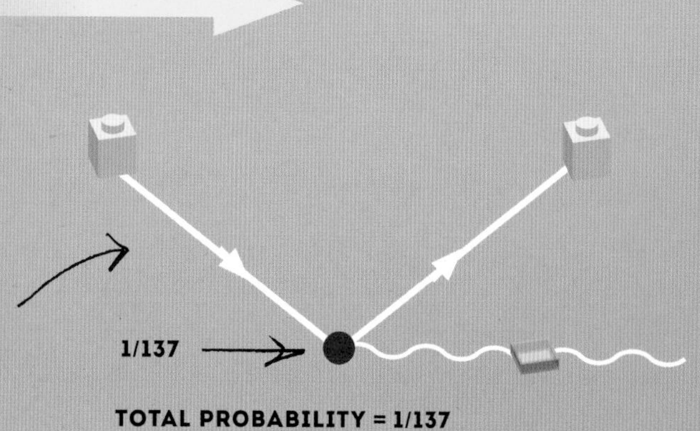

Flip a plastic brick and the chance it will land studs up is about ½. Flip it again and the likelihood of it landing studs up twice in a row decreases to $(½)^2 = ¼$ as there are $2^2 = 4$ possible outcomes.

Flip it again and the chance of landing studs up three times in a row reduces even further to $(½)^3 = 1/8$.

ELECTROMAGNETISM

Electromagnetism is the combined force of electricity and magnetism. A moving electric charge creates a magnetic field, the principle behind electromagnets. A magnetic field will therefore influence the path of a moving electric charge.

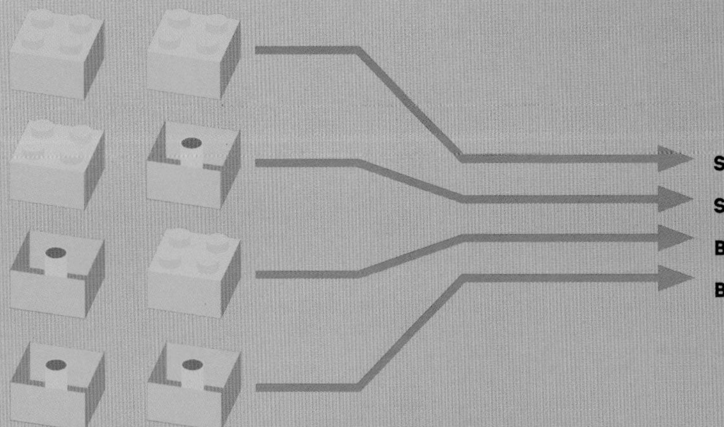

LOOP DIAGRAMS

We can draw another Feynman diagram that starts with an electron and ends with an electron and a photon. However, this diagram has not one but three vertices: two emitting a photon and one absorbing one – one photon is emitted and absorbed in a loop.

If the likelihood of each electromagnetic vertex is 1/137 the likelihood of three vertices is $(1/137)^3 = 1/2{,}571{,}353$. The more photons a particle is expected to emit or absorb, the less likely it becomes it will do so.

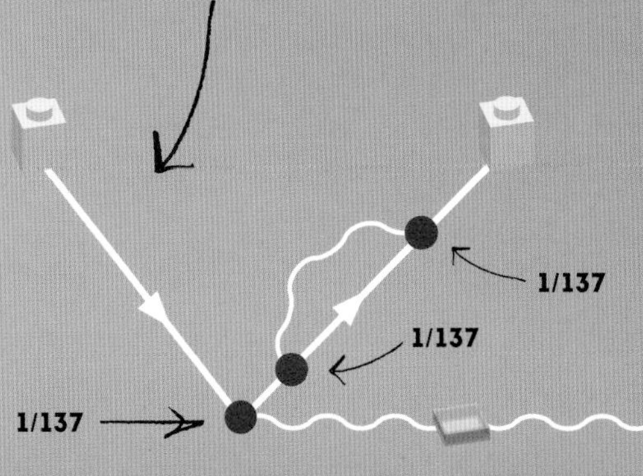

TOTAL PROBABILITY = $1/137^3$ = 1/2571353

The likelihood of these loop diagrams vanishes the more complex they become, leaving the most basic tree-level diagrams the most likely for any given incoming and outgoing particles. We will therefore only focus on these basic tree-level diagrams in the rest of the book.

Symmetries can be used to create and design new Feynman diagrams

SYMMETRIES AND FEYNMAN DIAGRAMS

The conservation of energy and momentum brings with it symmetry of space and time. The result is that we should not care about the direction of lines on a Feynman diagram in either space or time. It is legitimate to rotate them around the vertex. The only exception to this rule is that lines cannot *all* be to the left-hand side of the vertex or all to the right-hand side of the vertex. If they are, this implies either creation or destruction of energy which is not allowed! We always need a before and after story.

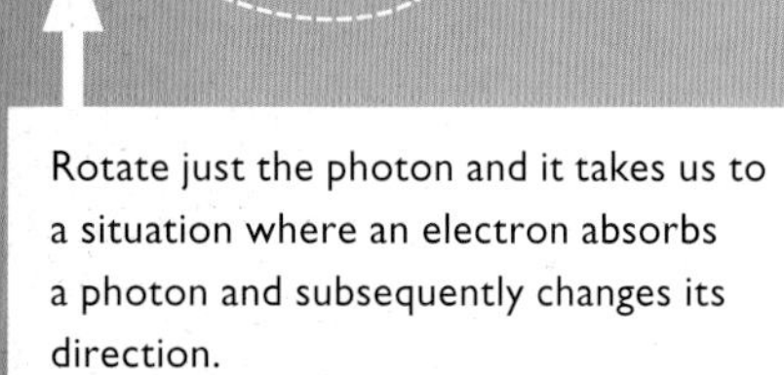

Rotate just the photon and it takes us to a situation where an electron absorbs a photon and subsequently changes its direction.

This diagram is forbidden by the laws of Nature as it shows particles (electron, antielectron and photon) being created from nothing.

ANTIMATTER

If we rotate either of the electron lines we get a strange scenario where the fermion line is now pointing backwards in time. Such strange things were first seen by Paul Dirac when he married quantum mechanics and Special Relativity. It is of course not a time-travelling particle of matter, but is in fact antimatter.

A fermion line pointing backwards in time represents an antifermion, in this case the antielectron, travelling forward in time. This shows a symmetry in the way electrons and antielectrons interact via the electromagnetic force with photons.

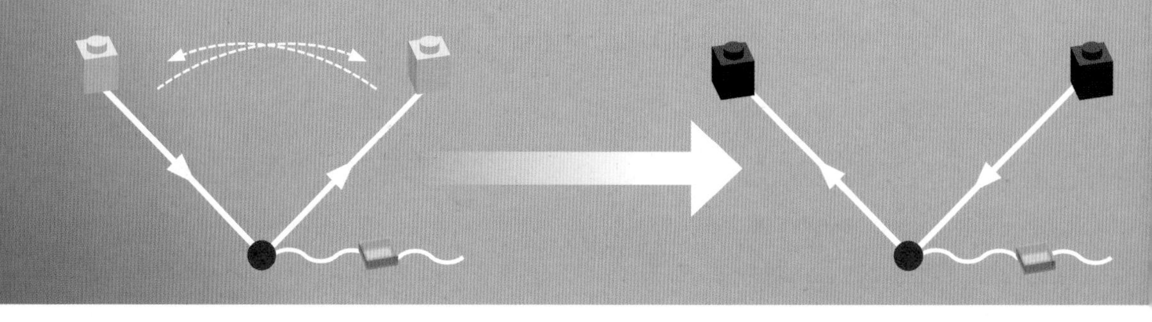

ANNIHILATION

If we rotate the electron from later time we get a scenario where an electron and antielectron meet and produce a photon. This is a legitimate interaction and is the fate of many particle–antiparticle pairs – annihilation. Whenever a particle meets its opposite number they destroy each other and create pure energy in the form of a boson.

SYMMETRIES

A symmetry is some transformation which leaves an object unchanged. The most familiar symmetries are rotation and reflection of shapes, but symmetries also exist in the mathematics of the Standard Model.

PAIR PRODUCTION

If we rotate both the earlier electron and the photon a story emerges of a photon giving birth to an electron and antielectron. This is known as pair production and is a legitimate thing which bosons do, giving up their energy to create fermions. Of course this interaction is only possible if the photon has energy enough to create the mass of both particles through $E = mc^2$. With an exchange rate of c^2 (speed of light squared) that is a lot of E (energy) to make a small m (mass).

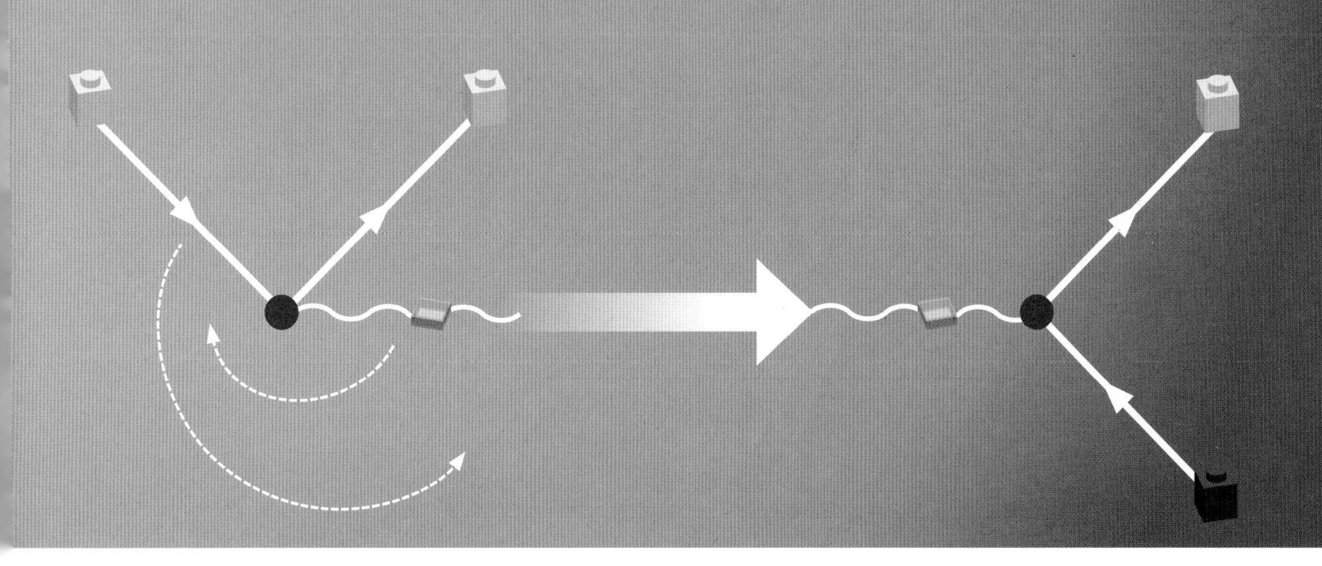

Particles can be detected through their electromagnetic interactions with the material around them

DETECTING PARTICLES

Particles are far too small to see in the same way as we see the world around us with light. The wavelength of light cannot be made small enough to notice a particle. We can see the studs on a plastic brick because of millions of photons scattering off the surface and into our eye.

To see small structures we need smaller things to fire at them, but the particles are by definition the smallest things, so how do we see them? Well we can't in this traditional way. Instead we infer their presence from their electromagnetic interactions with electrons swarming in atoms around them.

1

LUMINESCENCE

Luminescence is the process where an electric charge excites an electron to a higher energy which it loses some time later by emitting light. The light is then collected and its intensity measured.

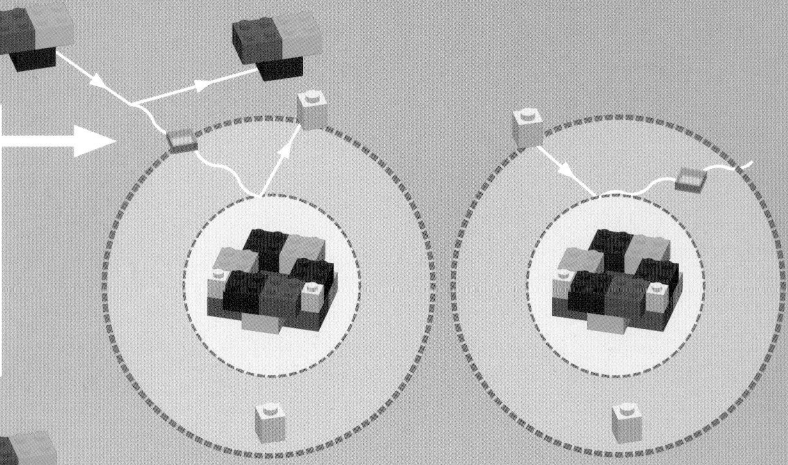

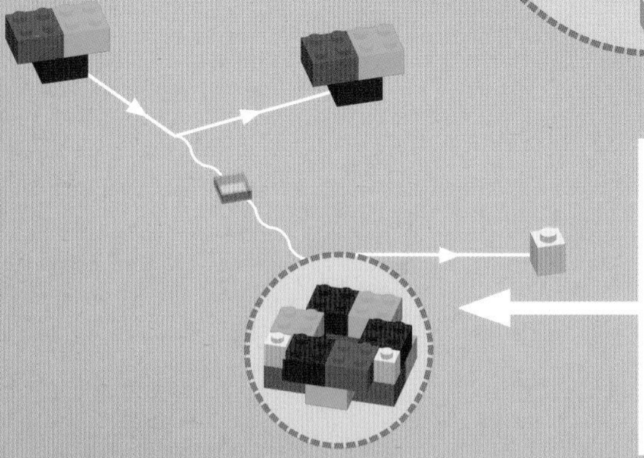

2

IONISATION

Ionisation is the process where an electric charge knocks an electron clean out of its host atom. These electrons can then be collected and counted.

By detecting the light emitted in luminescence or collecting the electrons emitted in ionisation we are able to build up a picture of where a particle has been, like the wake of water behind a speedboat or the contrails of an airplane's engine. The amount of light or number of electrons collected gives us information on the energy of the particles.

NEUTRINOS AND NEUTRONS

Electrically neutral particles like the neutrino and neutron are even further from direct sight, and again can only be observed in two ways:

1

ELASTIC SCATTERING

Neutral particles can knock charged particles clean out of a material. Neutrons were discovered because they knocked protons (hydrogen nuclei) out of hydrogen-rich paraffin wax.

PROTON

NEUTRON

PARAFFIN WAX nCH_2

NEUTRON

The unseen antineutrino transforms into an observable electrically charged antielectron.

ANTINEUTRINO

ANTIELECTRON

W^+

2

CHARGED CURRENT QUASI ELASTIC

Neutral neutrinos may also show their presence by transforming into charged particles through the weak force interactions. These charged particles are then observed. Neutrinos were discovered in this way.

Two quarks remain unchanged while one up quark is transformed by a weak force interaction to a down quark.

Particles with electric charge can shock vapour to form liquid droplets or emit light, making them visible

CLOUD AND BUBBLE CHAMBERS

Ever put a mint in a bottle of cola to watch the erupting reaction? The cola is super-saturated, having dissolved far more carbon dioxide gas than is comfortable, and the gas wants out. It remains in solution until there is some external trigger – the mint.

Some early particle physics detectors used this same principle. Instead of super-saturated solutions (although beer was tried at one point) these detectors used super-cooled vapours, with a temperature below their point of condensation, or super-heated liquids, at a temperature above their boiling point. If an electrically charged particle barges into either, the resulting ionisation provides a trigger for either the super-cooled vapour to form liquid droplets or for the super-heated liquid to form a vapour. Along its path the particle continues to ionise atoms around it which leaves a clear track that can be photographed. These detectors, called cloud and bubble chambers, respectively, were scaled to huge sizes and used to detect naturally occurring cosmic rays and particles accelerated by machine.

CHERENKOV RADIATION

Einstein's theory of special relativity states clearly that nothing can travel faster than light through empty space. Light does travel slower through different materials, slowed down thanks to interactions with electrons in the atoms it is made from.

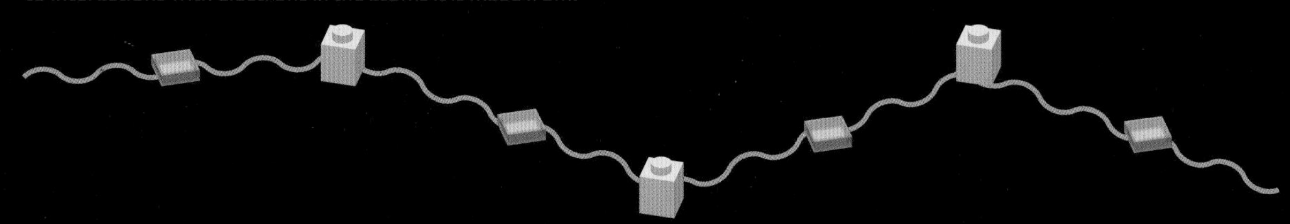

It is therefore not against the laws of physics for an electrically charged particle to be able to travel faster than light would in a material. When they do they emit light known as Cherenkov radiation. This is an optical version of a sonic boom, where a loud explosion is heard upon breaking the sound barrier.

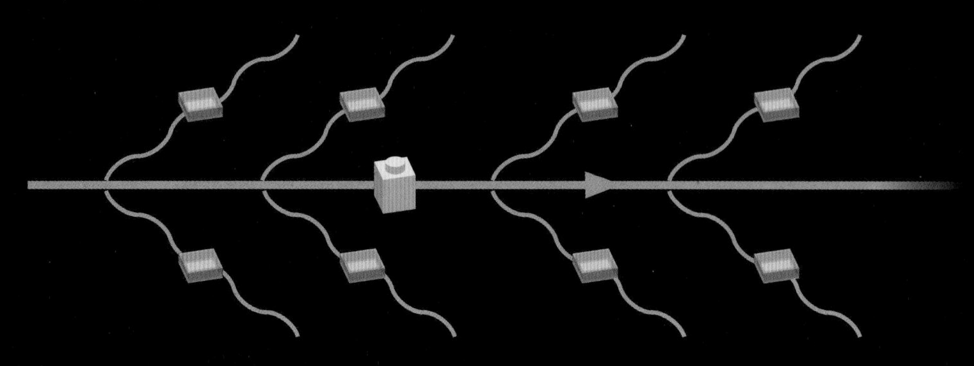

Select different particles of similar energy and lighter particles will be travelling faster than heavier particles; a golf ball will leave the hand faster than a football if thrown with the same force. If particles with similar energy are directed into a material it is possible that the low-mass particles are travelling fast enough to create Cherenkov radiation, while the more massive particles might not as they are travelling too slowly and so go unnoticed. By carefully selecting the material, detectors can be built in which only certain particles emit Cherenkov radiation. These Cherenkov counters allow accurate measurements of specific particles to be made.

The super-heated liquids in a bubble chamber change state into bubbles of gas when a charged particle passes through. The photograph below shows the trail of ionised atoms around which the bubbles formed. By measuring the curvature and energy loss of the particles as they travelled through the chamber, physicists could determine their energy, momentum and mass, discovering a host of new particles.

Modern particle detectors use different technologies in onion-like shells to detect all types of particle

MODERN PARTICLE DETECTORS

Modern particle detectors at particle accelerators have grown humongous in proportions to contain the huge amount of energetic debris from particle collisions. The discovery machines at the Large Hadron Collider at CERN are the largest and most advanced yet. They make use of almost every technology that has been developed in a multitude of ways to build up layers like an onion.

MUON TRACKER

SEEN: MUON

Muons do not feel the strong force and are heavy enough to steam train their way through the electromagnetic calorimeter which means they usually fly straight out of the detector. Catching one last glimpse of them as they leave are trackers, which have a reduced resolution compared with those in the centre of the detector.

UNSEEN: NEUTRINO

Neutrinos, having no electric charge and interacting so rarely, through the weak force only, do not get 'seen' at all. Instead their presence is calculated by assuming momentum is conserved and if there is some missing then it must have been taken by a neutrino.

Calorimeters turn the huge amount of energy particles have into showers of many more particles, thanks to $E = mc^2$. Each of these new particles has a small enough energy that they can be contained. Add their energies up and you get the energy of the original particle.

The hadronic calorimeter determines the energy of hadron particles, those made from quarks. The showers that hadrons create are a consequence of strong force.

HADRONIC CALORIMETER

SEEN: hadrons e.g. protons and neutrons

UNSEEN: neutrino and muon

An electromagnetic calorimeter converts electrons and photons into electromagnetic showers of new photons, electrons and antielectrons.

ELECTROMAGNETIC CALORIMETER

SEEN: electrons and photons

UNSEEN: neutrino, muon and hadrons

INNER TRACKER

SEEN: particles with an electric charge

UNSEEN: electrically neutral particles

Inner trackers, an advanced version of your digital camera, map the paths of thousands of electrically charged particles close to particle collisions by collecting ionised electrons in tiny pixels. They trace the tracks of particles to an accuracy of millionths of a metre.

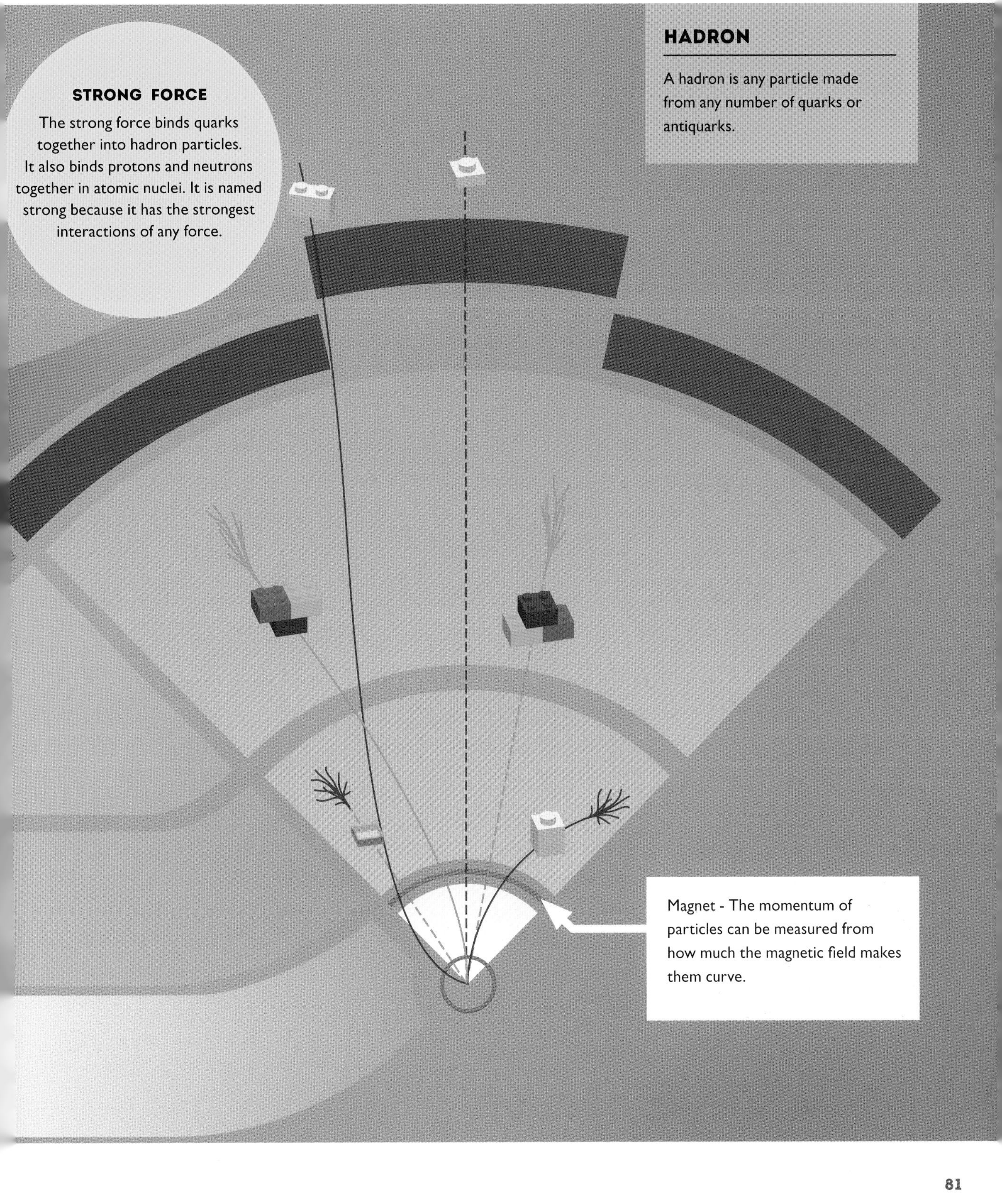

HADRON

A hadron is any particle made from any number of quarks or antiquarks.

STRONG FORCE

The strong force binds quarks together into hadron particles. It also binds protons and neutrons together in atomic nuclei. It is named strong because it has the strongest interactions of any force.

Magnet - The momentum of particles can be measured from how much the magnetic field makes them curve.

Particle mass can be determined by movement in magnetic fields or by the energy they lose when travelling through materials

MEASURING MASS

Identifying the mass of an object requires two things to be measured; the energy of a particle and its momentum. The energy of a particle is usually determined using a calorimeter as mentioned on the previous pages.

MOMENTUM

The momentum of a moving particle is determined by the amount it is bent by a magnetic field.

High-momentum particles will be bent a small amount.

Low-momentum particles will be bent a large amount.

CHARGE

Without knowledge of the energy of the particle, though, it is difficult to tell if the particle is a very fast moving low-mass particle or a slower moving high-mass particle. Another thing to consider is the electric charge on the particle. Positive electrically charged particles will curve in the opposite direction to negative electrically charged ones.

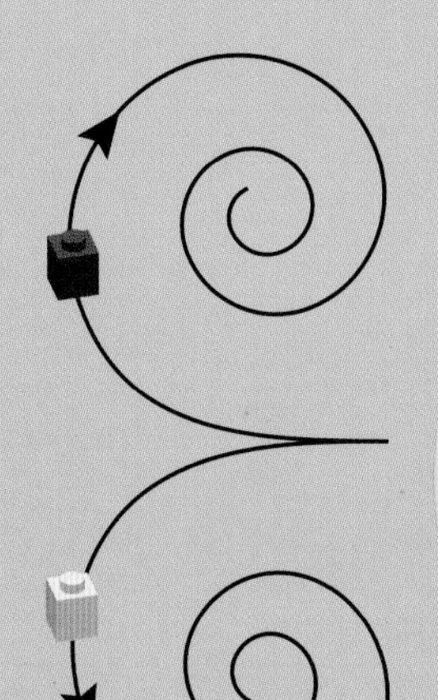

ELECTRIC CHARGE

The electric charge of a particle defines how it interacts with other particles through the electromagnetic force. Electric charge of particles can either be positive or negative and opposite charges attract while like charges repel.

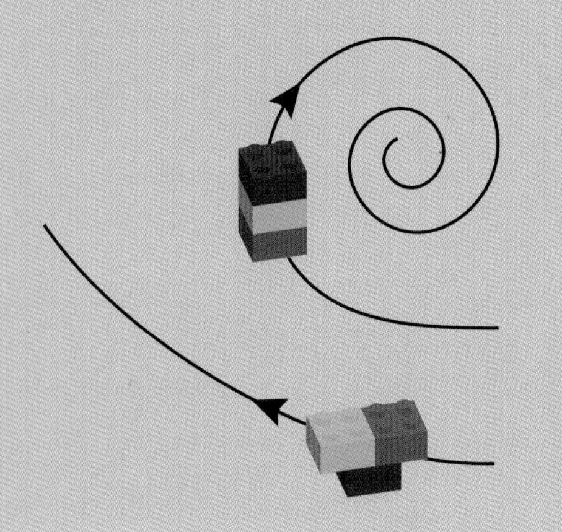

The larger the charge of a particle, however, the more it bends in a magnetic field. A single unit of either positive or negative charge is usually assumed, as this is the case for most particles.

BREMSSTRAHLUNG

Another method of determining the mass of a particle, used in early experiments, looks at the light which is emitted as an electrically charged particle passes through a material. When a particle passes by an atom its movement is affected by the atom's electric potential; it is given a hill to fall down. The particle changes direction and at the same time loses energy by emitting photons. This emitted light is something known as bremsstrahlung, translated from German as breaking radiation. As lighter particles have their paths changed more easily, they emit more radiation and lose energy faster.

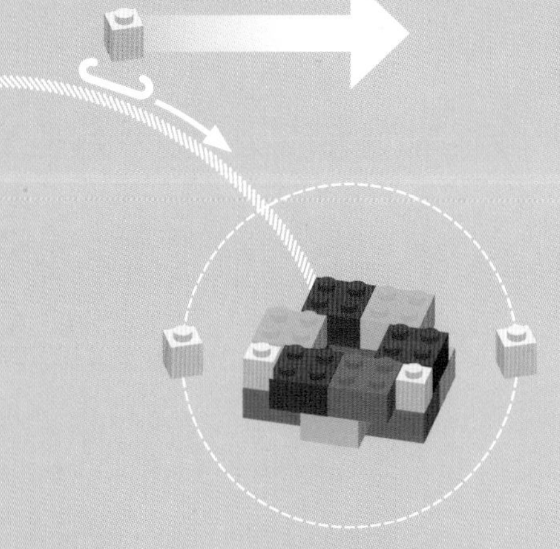

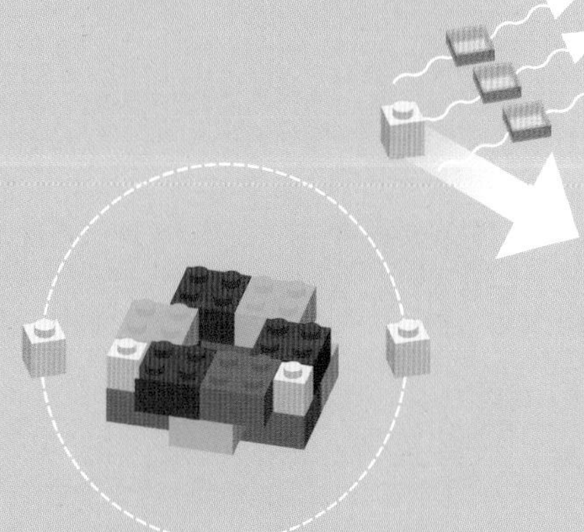

Larger mass particles lose less energy as they emit less bremsstrahlung. Measuring the tiny amount of bremsstrahlung identified the heavy muon in cosmic rays, the first of the second-generation particles to be discovered (see pages 88–9).

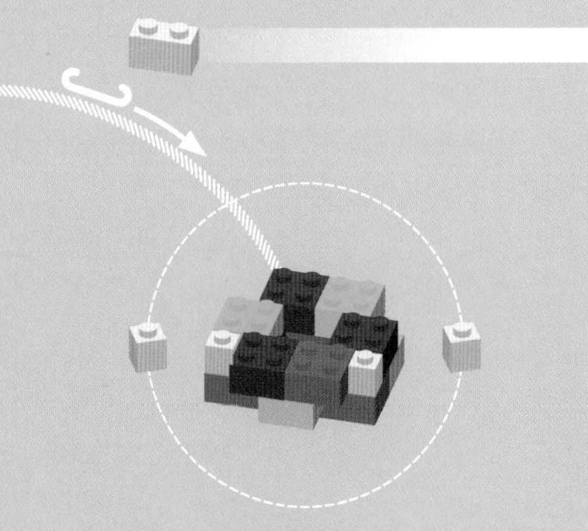

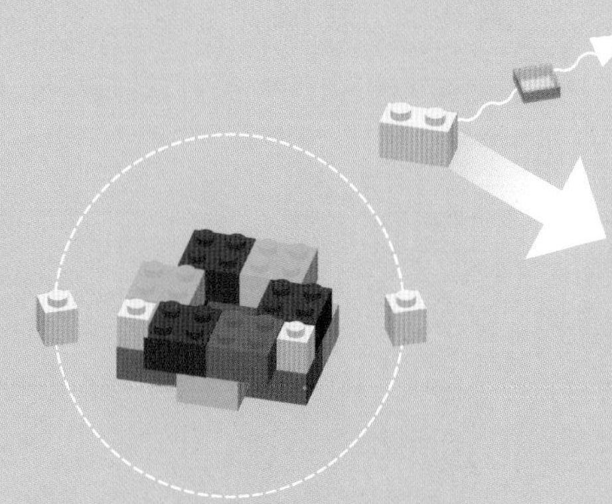

Radiation is emitted any time the path of a charged particle is altered either by an electric or a magnetic field. Particles accelerated around a synchrotron will lose energy in this way. This is a major limiting factor when trying to accelerate particles to ever higher energies. Some facilities, however, put this synchrotron radiation to good use as a probe to determine the structure of materials and biological tissue.

Spin determines whether a particle can build atoms or acts as a messenger for a force

QUANTUM SPIN

The deflection of charged particles in magnetic fields suggests that they behave like tiny magnets. Their magnetic behaviour is very much like that of spinning electrically charged balls. But thinking of particles as little spinning balls does not take us very far.

Electrically charged particles would have to spin faster than the speed of light if they were to generate the magnetic field observed.

In 1922, German scientists Otto Stern and Walther Gerlach investigated the deflection of silver atoms after they had passed through a strong magnetic field. If the spin of the atom were like a ball, then it and the magnetic poles produced could face in any random direction. This would show as a random spread of atoms in Stern and Gerlach's experiment.

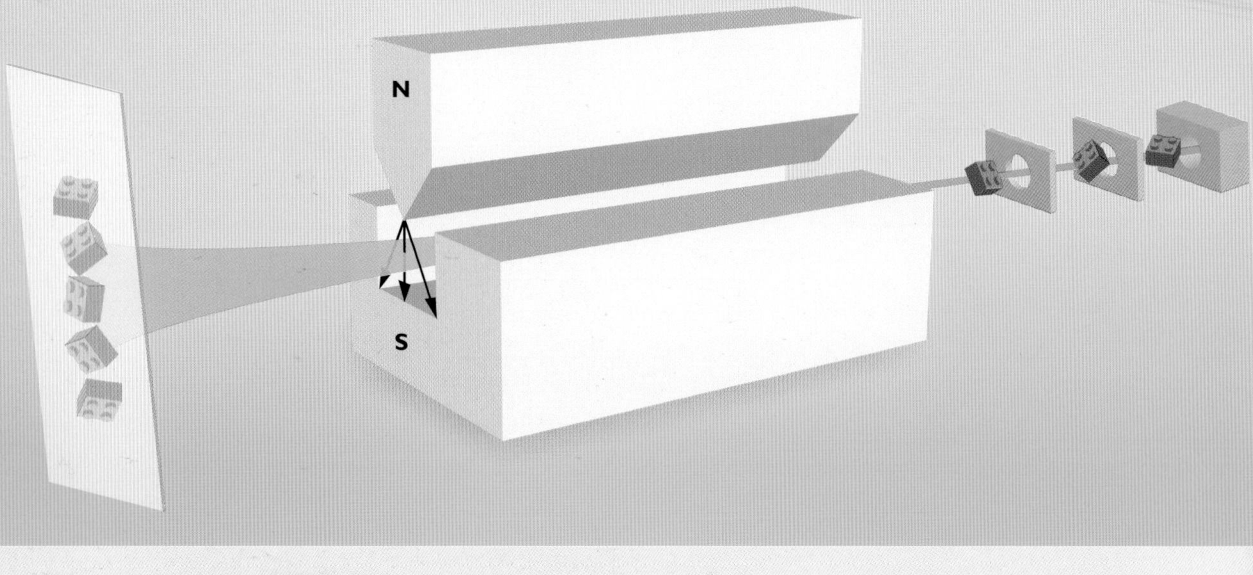

N

S

Instead, the atoms were either deflected up or down by the same amount, and not spread randomly at all. This showed that the magnetic poles of the atom, and by assumption the 'spin' of the atoms, could only have two quantized values for spin and it did not depend on the orientation of the particles in space. This quantum spin had to be an internal property of particles and atoms.

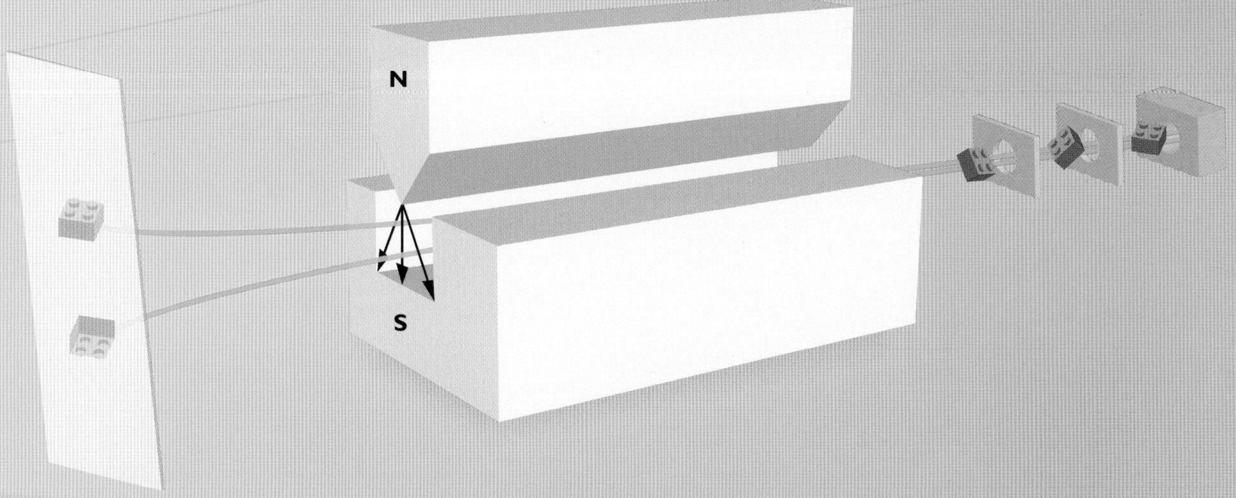

N

S

Two cars travelling on a highway can both be travelling at a speed of 100 km/h, but if one is travelling north and the other south, then they will end up in very different destinations. To distinguish the two we focus instead on their velocity which not only accounts for the magnitude of speed but also their direction. With one direction defined as positive, one car will be travelling at +100 km/h while the other travels at -100 km/h.

Different particles take different magnitudes of spin; fermions ½, Higgs boson 0, and all other bosons 1. Like the velocities of cars, depending on the direction of their spins they can take values which can be different by a value of 1.

This means that any fermion can take one of two values for their spin, either +½ or -½. A Higgs boson can take only one value for its spin of 0.

Other bosons can be found in three orientations; +1, 0 and -1. While the plus and minus one makes sense in the car analogy above, where does zero fit in? Well if we are in a car driving alongside another and matching its speed, then to us the car would not seem to be travelling north or south, its velocity in those directions compared with us would be zero. This is the situation for bosons with spin 0. The only problem is that the spin 0 option is not open to massless photons or gluons because they can never be matched for speed as they are travelling at the speed of light.

Radioactive chemical elements led scientists into the subatomic world

RADIOACTIVITY

Radioactivity, a word coined by Marie Curie, is the name given to any radiation actively emitted from an atom. There are three types of radioactivity and their names were given by New Zealand-born Ernest Rutherford as simply the first three letters of the Greek alphabet; α, β and γ.

ALPHA

Alpha (α) is low-penetrating radiation made from a stream of helium-4 nuclei ejected from the nucleus of heavy elements. As helium nuclei are bulky and carry a large plus two charge, they readily lose energy as they pass through a material, usually stopped by a sheet of paper. The radiation lends it name to the fusion process involving helium-4 nuclei in the Sun (see pages 50–1). Alpha radiation allows a heavy element to shed a large amount of mass, making it much lighter and usually more stable overall. Because of this, alpha decay is predominantly seen in Uranium and heavier elements, while the transformative beta decay can be seen in lighter radioactive elements such as Technetium.

Marie Curie was a key founder of nuclear physics, coining the term radioactivity for the radiation emitted from the nucleus of heavy unstable elements. She is also one of the few women truly recognized in their lifetime for their important scientific work, with a Nobel Prize in both Physics (1903) and Chemistry (1911).

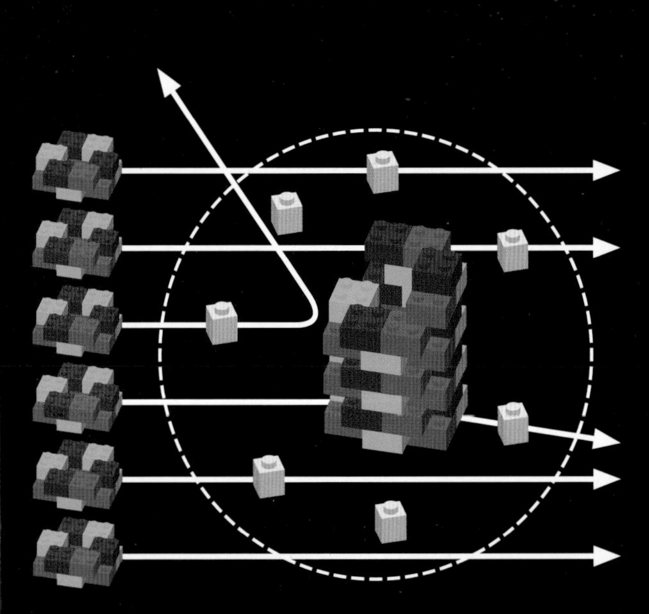

Alpha radiation was used by Hans Geiger and Edward Marsden in their discovery of the atomic nucleus. After pointing alpha radiation at thin gold foil, they found that some of the alpha particles bounced right back in the direction they were originally fired. Such dramatic rebounds could only be explained if there were a dense and electrically charged region inside the atom, the nucleus.

BETA

Beta (β) is more penetrating, making it through paper but stopped by thin metal foil. It was found to be made from fast-moving electrons. They are emitted when there is a change from neutron to proton in the atomic nucleus. There is more on this later in chapter 5.

β

GAMMA

Gamma (γ), the third and yet more penetrating radiation, is made of high-energy photons which are emitted when protons and neutrons rearrange themselves to sit more stably in the nucleus. It takes a few centimetres of lead to stop.

γ

A muon is a heavier version of the electron; antimuons and antielectrons are the respective antiparticles

THE ANTIELECTRON AND THE MUON

In the 1930s, particle physicists' eyes turned to the skies and the high energy cosmic radiation raining down upon us (see pages 94–5). Among the menagerie of particles being discovered there were two which behaved quite differently. One had been predicted by the mathematics of the evolving quantum theory of particles, and confirmed the existence of a new class of particle. The second came literally out of the blue, taking everyone by surprise, to change forever the number of fundamental building blocks from which Nature is constructed.

ANTIELECTRON

The antielectron, predicted by Paul Dirac in his equations explaining quantum particles, was discovered by American Carl Anderson in cosmic rays. He placed a magnet around a cloud chamber and observed a particle with the same mass as an electron curving in the opposite direction to the one expected for a negatively charged particle. The mass of this positively charged antielectron was determined by the amount of energy it lost, thanks to bremsstrahlung, as it passed through a sheet of metal – the particle curving more afterwards as it had slowed down after losing energy.

BREMSSTRAHLUNG

Bremsstrahlung, or braking radiation in English, is the light emitted when the path of a particle is altered when passing by an atom. All particles emit bremsstrahlung when passing through matter, the lighter the particle the greater the amount of bremsstrahlung emitted and the quicker they lose their energy.

It was not long after the discovery of the antielectron that events were seen in which electron and antielectron were produced in a pair by an otherwise unseen photon.

MUON

Later Anderson and colleague Seth Neddermeyer observed new particles which did not lose as much energy when passing through metal plates. Unable to explain this new particle, theoretical physicists quietly invoked desperate remedies such as there being 'red and green' types of electron.

In the end it just turned out to be a new particle altogether – the muon. These penetrating particles were not losing energy through bremsstrahlung because they were far more massive than electrons. The positive electrically charged antimuon was also seen in cosmic rays.

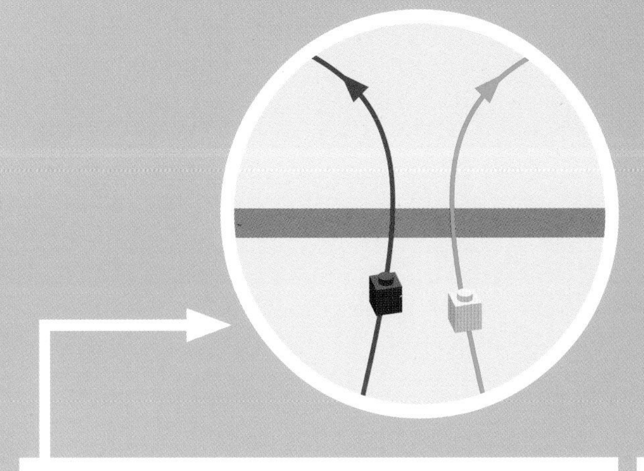

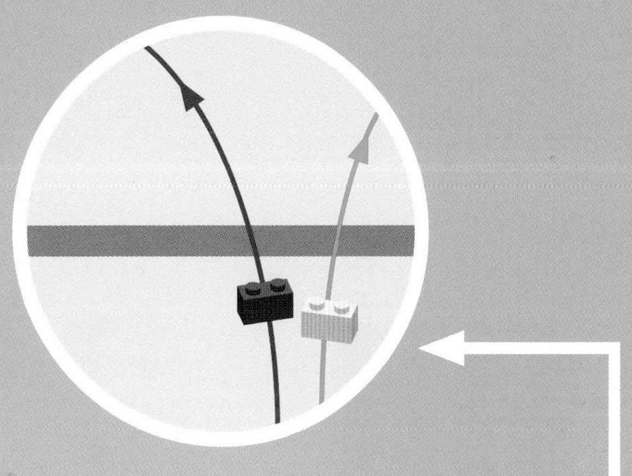

The paths of electrons became more curved due to losing energy after passing through a metal plate.

The curvature of muon paths did not change much at all after passing through metal plates which demonstrated that they had not lost much energy.

We know today that the muon is the heavier cousin of the electron, around 206 times heavier at 105 MeV/c^2. Just as soon as particle physics seemed to be wrapped up with everything made from protons, neutrons and electrons, along came the muon. A discovery so unexpected that Nobel Prize-winning physicist Isidor Isaac Rabi exclaimed "Who ordered that?"

Carl Anderson shows off his cloud chamber which was used in the discovery of the antielectron.

ELECTROMAGNETISM AND QED SUMMARY

The force of electromagnetism is the lens through which we observe and identify particles, thanks to the infinite range of the photon.

Opposite electrically charged particles attract while like charged particles repel one another through the electromagnetic force.

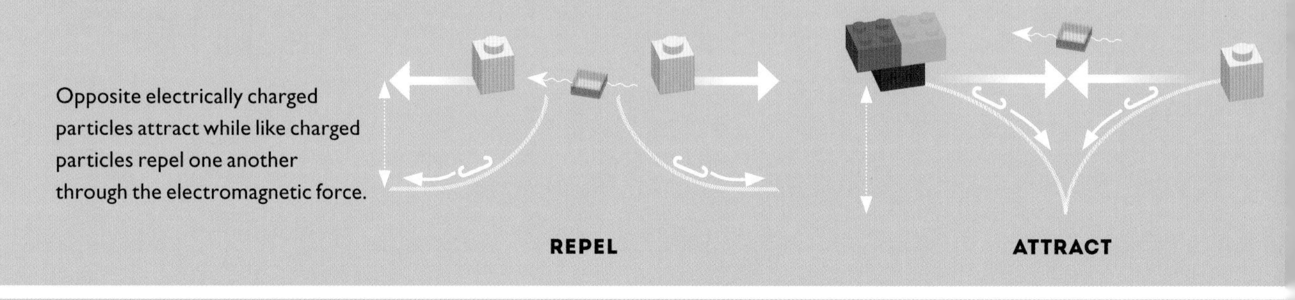

An electrically neutral atom is composed of a positive electrically charged atomic nucleus containing a certain number of protons surrounded by an equal amount of negative electrically charged electrons.

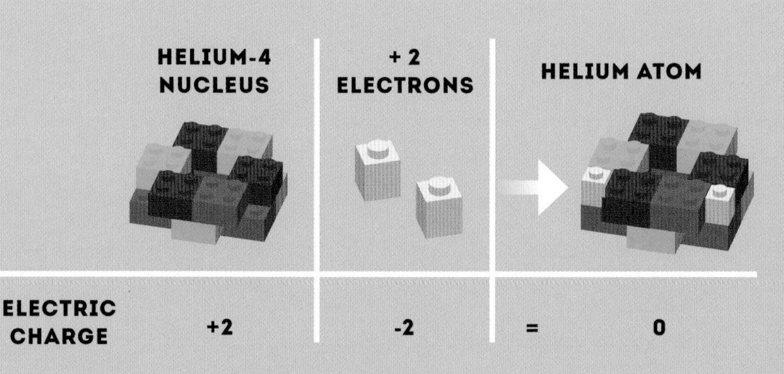

The electromagnetic force is exchanged between particles with electric charge by the photon and all particle interactions can be drawn in cartoons called Feynman diagrams.

Straight lines represent fermions with an arrow to show their direction.

Plastic brick to show the type of fermion (symbols are used in standard Feynman diagrams).

The vertical axis represents the position of particles in space.

The vertex is the point at which these lines meet and exchanges occur.

The horizontal direction represents the passing of time, from earlier times on the left to later times on the right.

Wiggly lines represent the force-carrying boson (a photon in this example).

Particles and antiparticles with electric charge can annihilate one another to form photons and a photon can create a particle–antiparticle pair.

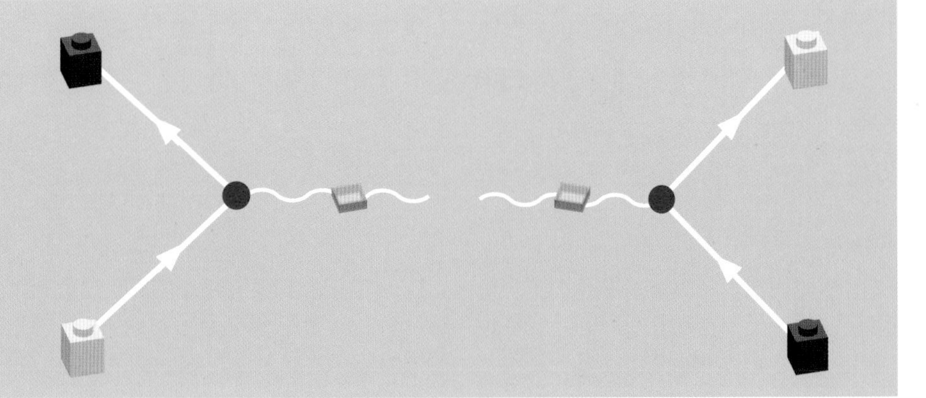

Particles with an electrical charge can be accelerated to high energies on a wave of electric potential using the idea of attraction and repulsion.

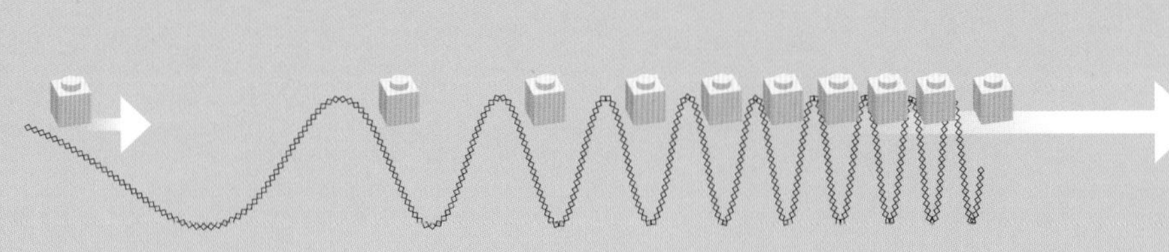

All particles with an electric charge are deflected by magnetic fields in opposite directions for opposing electric charges.

As particles are accelerated by attraction or repulsion from an atomic nucleus they emit light known as bremsstrahlung – the lighter the particle, the greater the acceleration and the more energy is lost through this emission of light.

Particles behave very differently to bricks when travelling from one place to another because of internal pedometers which tick by as they move. The electromagnetic force results from a symmetry between internal pedometers of particles, one which means that particles do not care about the original setting of pedometers but instead just their relative change.

Mesons carry the strong force which keeps protons together in atomic nuclei

YUKAWA'S MESONS – PIONS

Without the strong force, nuclei, and therefore atoms, could not exist – the positive electrically charged protons would just push each other apart. In this chapter we will discover more specimens in the zoo of particles which are all bound together with the strong force. This will tell us more about how the strong force influences the lives of all matter around us.

Just like photons exchanging the electromagnetic force, Japanese physicist Hideki Yukawa predicted in 1935 that there must exist particles in the nucleus exchanging a strong nuclear force between protons and neutrons.

In our brick model it is the studs which bind the protons and neutrons together. In a real nucleus they are bound by exchanging Yukawa's new particles called mesons.

Yukawa named the particles mesons, after the Greek *mesos* meaning middle, as he predicted them to have a mass between that of the light electron and heavier proton. Today however, the word meson is used to describe any particle which is made up from a quark–antiquark pair.

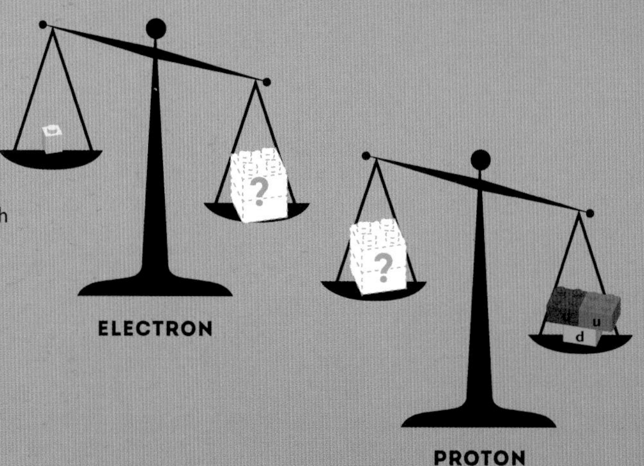

The first sighting of these mesons, later named pions and symbolised by the Greek letter π, was in their decay to muons, which showed that they had a slightly larger mass.

When pions are exchanged between protons and neutrons they mix them up, changing protons to neutrons and vice versa, without changing the overall number of each. There are several ways that they do this.

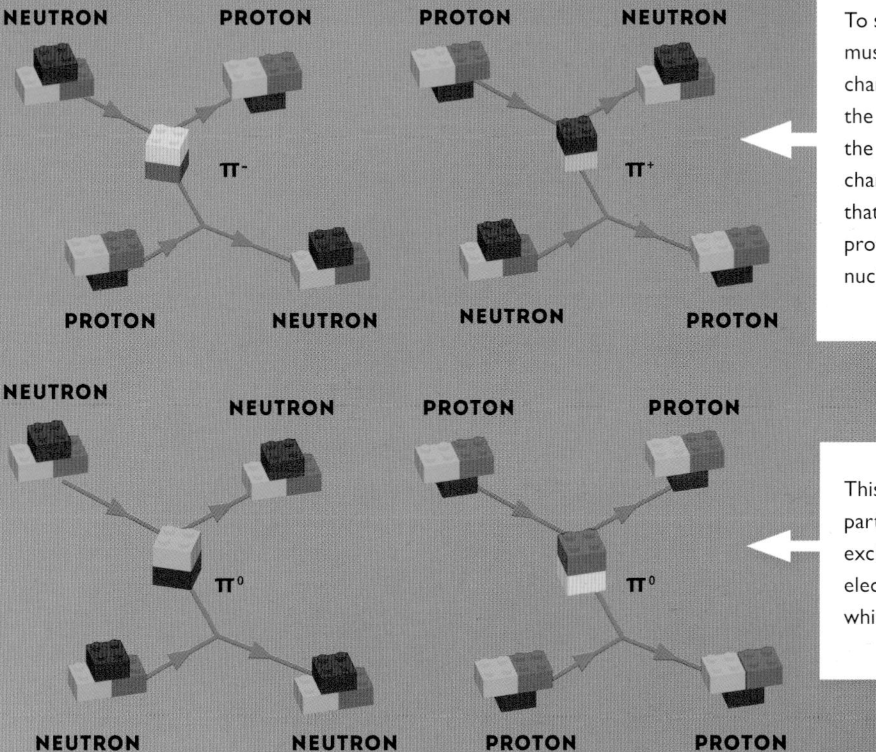

To swap neutron for proton, a pion must take away a negative electric charge to balance the positive of the proton – introducing the π^-. For the opposite we get its positively charged antiparticle, the π^+. Note that both start and end with one proton and one neutron, so the nucleus remains unchanged.

This exchange does not swap particles around but only exchanges the strong force. It is an electrically neutral pion, the π^0, which is its own antiparticle.

It took the advent of particle accelerators before the neutral pion, π^0, was confirmed to exist. Experiments counted pairs of photons and added up their energy to calculate the mass of the particle from which they came. Pions have masses of 140 MeV for the π^+ and π^-, and 135 MeV for the π^0, truly a meson, in between the electron at 0.511 MeV and proton at 938 MeV, as predicted.

Cosmic rays arrive at the Earth from high-energy astronomical source

COSMIC RADIATION

Yukawa's meson and other new and exotic forms of particle were discovered, like the nucleus before them, thanks to a new form of radiation. In 1912, Austrian Victor Hess detected increasing amounts of radiation on a balloon flight as he flew higher. A new radiation was discovered coming from the cosmos. Every second of the day showers of energetic particles rain down upon us after being created kilometres above our heads. These cosmic ray showers are born when a particle exchanges huge amounts of energy in interactions with atomic nuclei in the Earth's atmosphere to create a mass of new and interesting particles in a cascade of decay.

Cosmic rays provided scientists with the first view of Nature at high energies. Today cosmic rays still provide us insight into the Universe at the highest energies, reaching far beyond the capability of the Large Hadron Collider. Understanding cosmic rays is important as they are one of the contributing factors to cancer, regularly breaking apart our DNA only for our cells to repair it. While we can't escape cosmic rays entirely, those of us who spend large amounts of time at high altitude, such as flight attendants and pilots, leave themselves at greater risk of developing cancer due to the higher rate of cosmic rays just as Victor Hess observed in 1912.

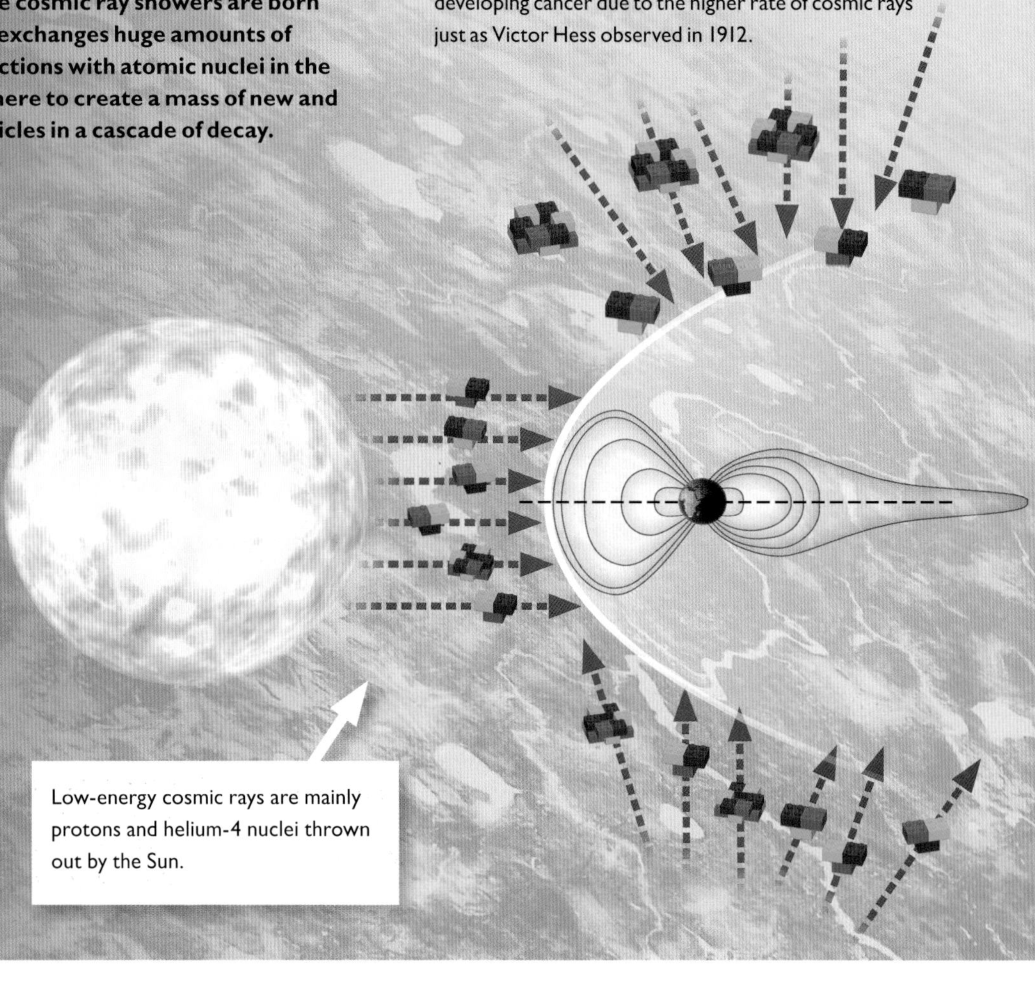

Low-energy cosmic rays are mainly protons and helium-4 nuclei thrown out by the Sun.

The origin of high-energy cosmic rays is an area of debate. Most are thought to come from the merging of supermassive stars or black holes. Some protons have been observed to pack the same energy as a tennis ball served by a professional. Both high- and low-energy cosmic rays create showers like the one shown below.

Protons make up 90% of cosmic rays which strike the upper atmosphere, 9% are helium-4 nuclei and just 1% are nuclei of heavier elements. Cosmic rays collide with the nuclei of gases in the Earth's atmosphere and their energy (E) is transferred into mass (m), creating a cacophony of new matter, thanks to the relationship $E=mc^2$.

Exotic particles are heavy and unstable and rarely make it to ground level. Most decay into lighter particles such as the muon, the heavier cousin of the electron.

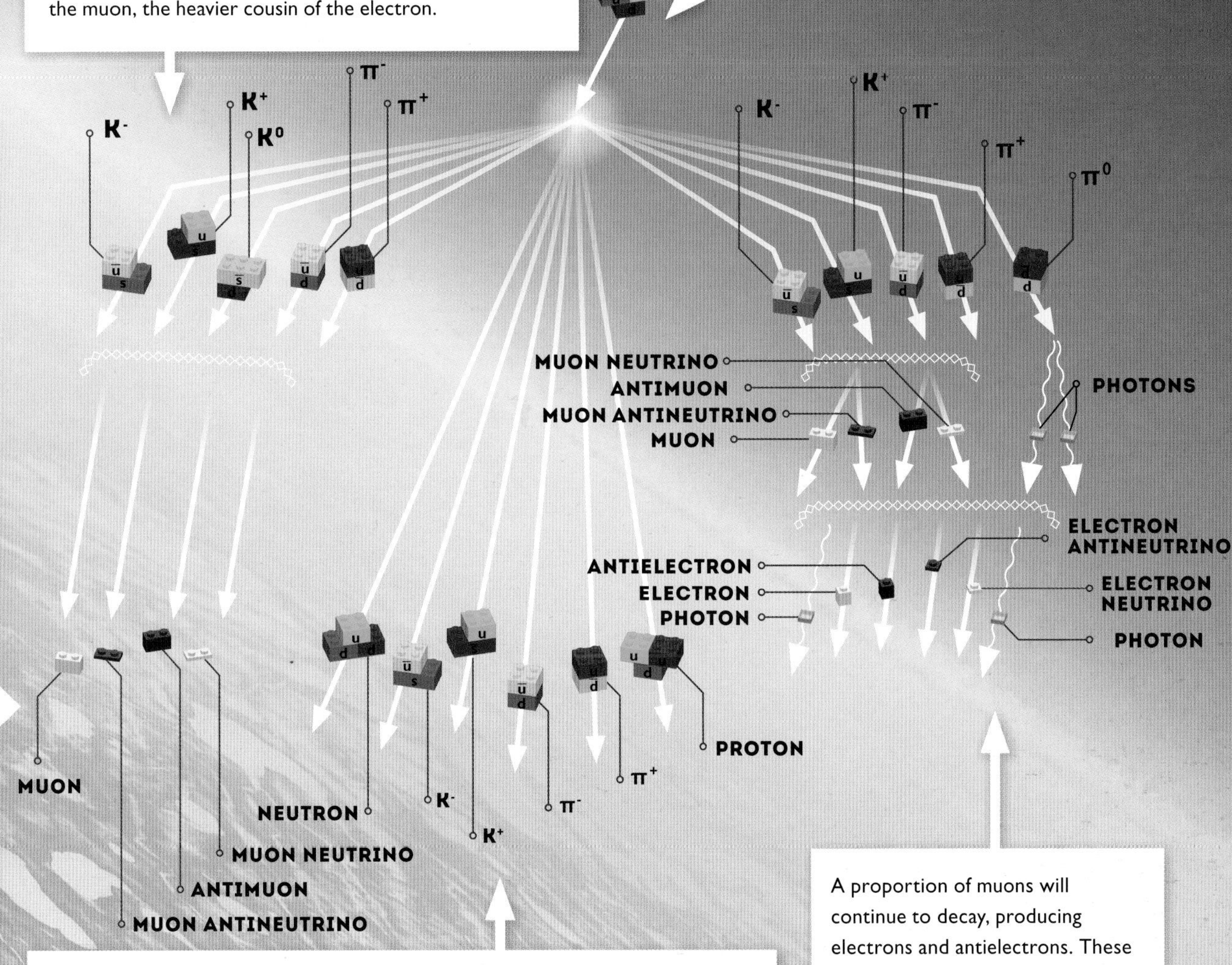

New exotic particles (different from the proton, neutron and electron) are created thanks to the strong nuclear force when the nuclei of atoms fragment in the collision. We will talk about these new exotic forms of matter in the following pages.

A proportion of muons will continue to decay, producing electrons and antielectrons. These result in showers of electrons, antielectrons and photons which are born from interactions of the electromagnetic force.

More mesons, as well as hyperons, can be found in cosmic rays

MORE MESONS AND HEAVY HYPERONS

Other mesons aside from pions were also seen in cosmic ray showers, behaving very strangely compared to other particles. The tau meson, kappa meson, and omega meson were soon found to have a similar mass and therefore to be different examples of the one and the same particle – the K meson, or kaon for short.

The particle passes through the lead without losing too much energy through bremsstrahlung (see chapter 3), proving that the particle produced is a muon and not an electron.

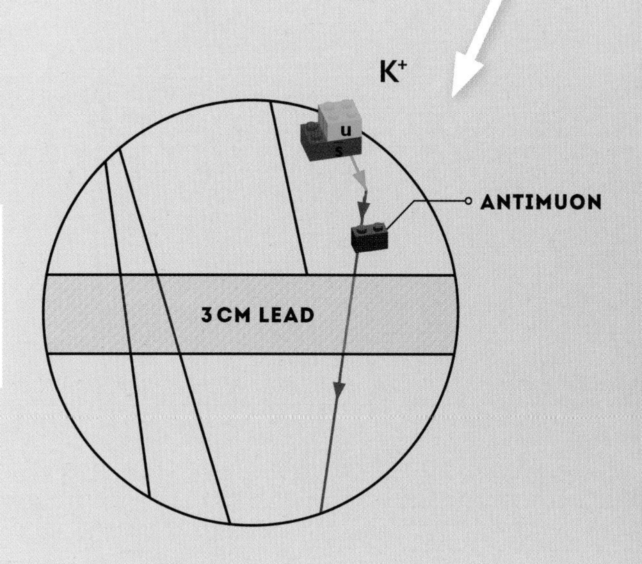

Like the pion there are positive and negative kaons which are each other's antiparticle and were first seen by their decay into muons.

K^- K^+

Unlike the neutral pion, which is its own antiparticle, there are two distinct neutral kaons, the K^0 and $\overline{K}^0$ identified through the different ways in which they decay – more on these later.

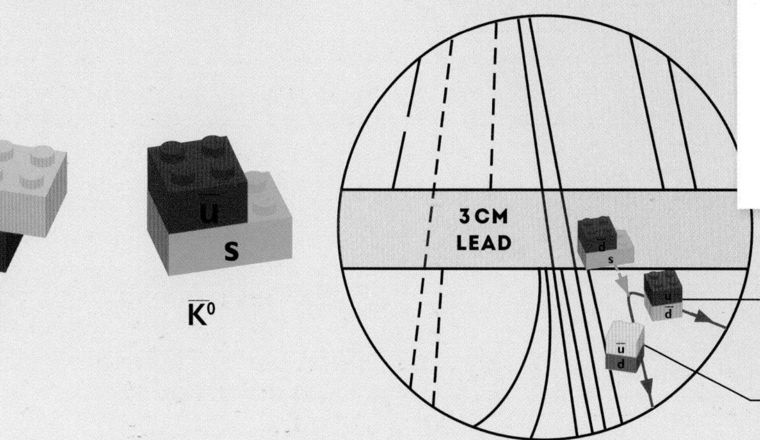

π^+

π^-

HYPERONS

The kaons were always seen created in partnership with other, much heavier, particles. They were given the name hyperons, as they were 'hyper' the mass of the nucleons (protons and neutrons) which were the previous heavyweight title holders. The tradition of naming particles with Greek letters stuck. The electrically neutral lambda hyperon, Λ^0, had no other partners.

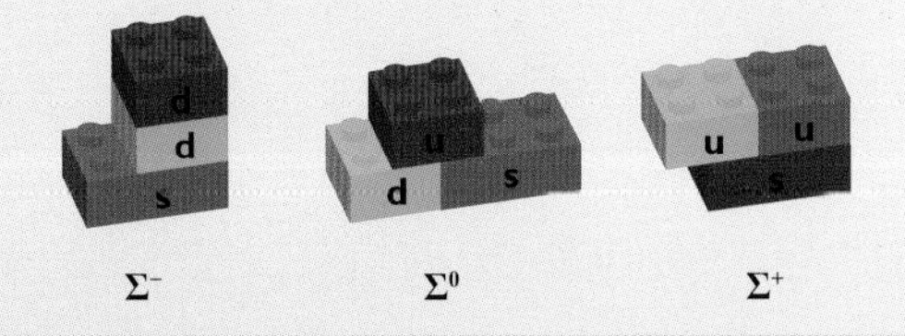

The sigma hyperon came with an electric charge of plus one Σ^+, minus one Σ^-, and electrically neutral, Σ^0.

STRANGENESS

Just as electrons and antielectrons are produced in pairs from a photon with zero electric charge, thus conserving electric charge, theoretical physicists Murray Gell-Mann and Kazuhiko Nishijima suggested there should be a new conserved quantity. This new conserved quantity was called strangeness and was conserved in the production of hyperons and kaons as they had opposite strangeness numbers – hyperons +1 and kaons -1.

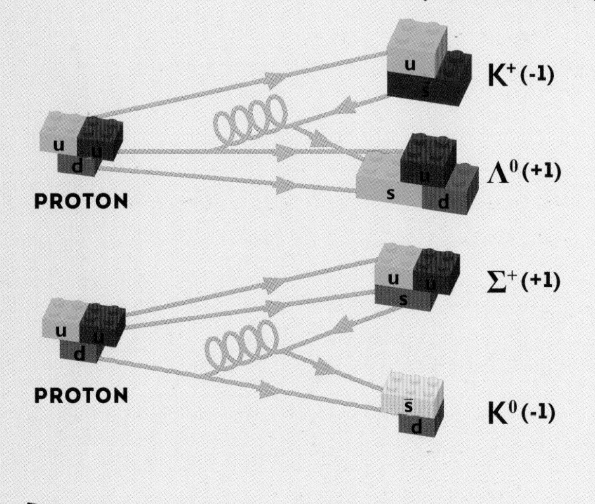

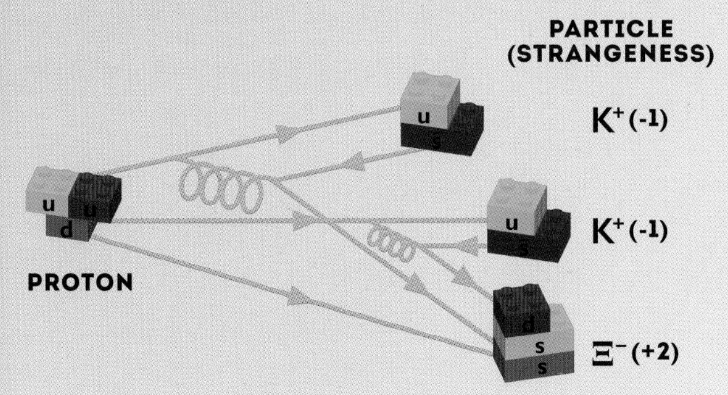

Xi hyperons, heavier still because they decayed into sigma hyperons, show themselves to have even more strangeness. They have strangeness of +2 as each of the two types, negative Ξ^- and neutral Ξ^0, were created alongside two kaons.

You can probably guess from the diagrams that the strangeness of a particle has to do with the number of strange quarks and antiquarks within it. Each strange quark adds one to the strangeness of a particle while an antiquark subtracts one.

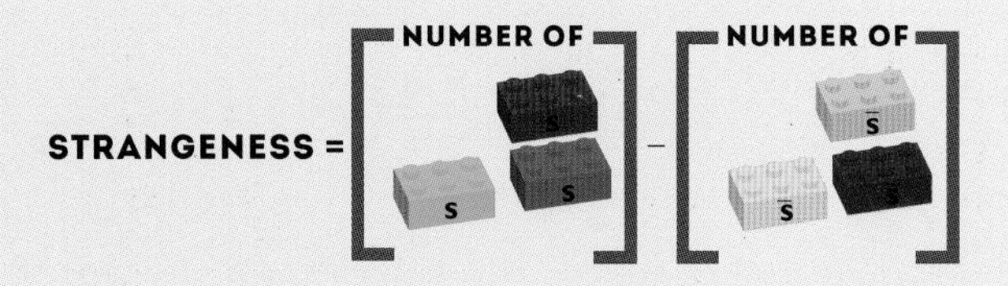

Particle accelerators enabled us to produce and detect antineutrons and antiprotons

ANTIPROTON AND ANTINEUTRON

In the 1950s and 60s, particle accelerators were getting larger, pushing particles to ever higher energies. The Bevatron (billions of eV synchrotron) at Lawrence Berkeley National Lab in the US smashed protons head on into each other. It was where the antiproton was first seen, earning Emilio Segrè and Owen Chamberlain the 1959 Nobel Prize in Physics.

The discovery of the antiproton showed that there could exist entire anti-atoms. Experiments at CERN today produce, trap and measure the properties of antihydrogen atoms.

PARTICLE ACCELERATORS

Particle accelerators are machines which use electric fields to boost the velocity of electrically charged particles.

P u u d

P u u d

$\bar{u}$ $\bar{u}$ $\bar{d}$ $\bar{P}$

P u u d

1 + 1 - 1 + 1 = +2

The antiproton is created when two protons try to rip each other apart in high-energy collisions. The huge amount of energy is transferred into the mass of new quarks and antiquarks which become bound by the strong force to produce new particles and antiparticles.

As with Feynman diagrams, look at this from left to right, in the direction of increasing time. Any arrow drawn going backwards in time, right to left, represents antiparticles.

$\bar{p}$ $\bar{u}$ $\bar{u}$ $\bar{d}$ — n: u d d

p: u u d — $\bar{n}$: $\bar{u}$ $\bar{d}$ $\bar{d}$

BARYON NUMBER 1 - 1 = 0

1 - 1 = 0

Just a year later the same machine, now smashing antiprotons into protons, discovered the antineutron. It seemed that every particle, not just the electron, has an antiparticle partner.

Looking at the creation of the antibaryons above, you might notice another symmetry. If all baryons had an associated baryon number +1 and all antibaryons a baryon number of -1 then the total before and after the interactions remains the same. Baryon numbers are conserved in all interactions be they strong force, weak force or electromagnetic.

ISOSPIN

There is one other symmetry which I haven't mentioned yet and that is called isospin. It groups the proton and neutron together into a pair which, as Yukawa predicted, are treated equally by the strong force. The number is related in mathematics only to the quantum spin of particles, which for fermions like protons and neutrons is ½. Protons have an isospin of +½ and neutrons -½, calculated from the number of up and down quarks and antiquarks they are made from.

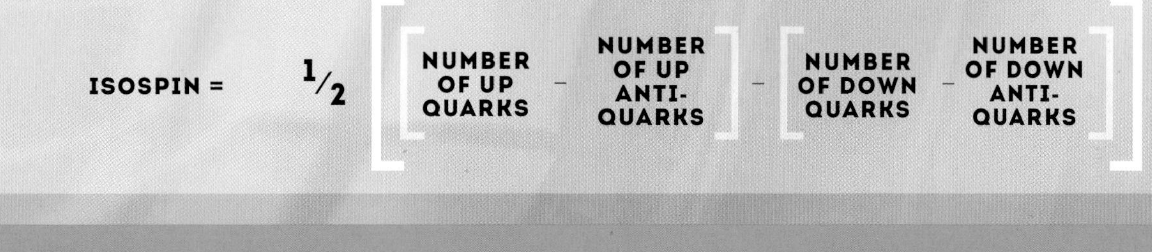

BARYONS

The proton, neutron, hyperons and their antiparticles are all heavier than the leptons and mesons and are collectively called baryons – from the Greek for heavy, *barys*. Baryons are also defined as being made from three quarks and antibaryons from three antiquarks. Mesons on the other hand are made from one quark and one antiquark.

Physicists noticed a pattern in the ever-growing zoo of new particles being discovered

PATTERN FINDING AND THE EIGHTFOLD WAY

Murray Gell-Mann not only identified the symmetry of strangeness but spotted a symmetry which explained the entire zoo of particles. Combining the idea of isospin and strangeness he developed a three-dimensional symmetry which not only placed the already-detected particles but also predicted new ones. The eightfold way was the first step to realizing this underlying structure to the zoo of particles seen at particle accelerators. While on the face of it particle physics seemed to be becoming ever more complicated, the eightfold way suggested a deeper simplicity yet to be uncovered.

The known baryons were grouped into a collection of eight called an octet and a collection of ten called a decuplet. The collection of eight were the protons, neutrons and hyperons we have discussed and they were laid out depending on their isospin and strangeness. The quantum spins of two quarks in each of these particles were facing in opposite directions and so overall they had the ½ spin of the third quark.

STRANGENESS

The strangeness of a particle is a number which represents the number of strange quarks or antiquarks within a hadron. Each strange quark subtracts 1 from to the strangeness number while each strange antiquark adds 1.

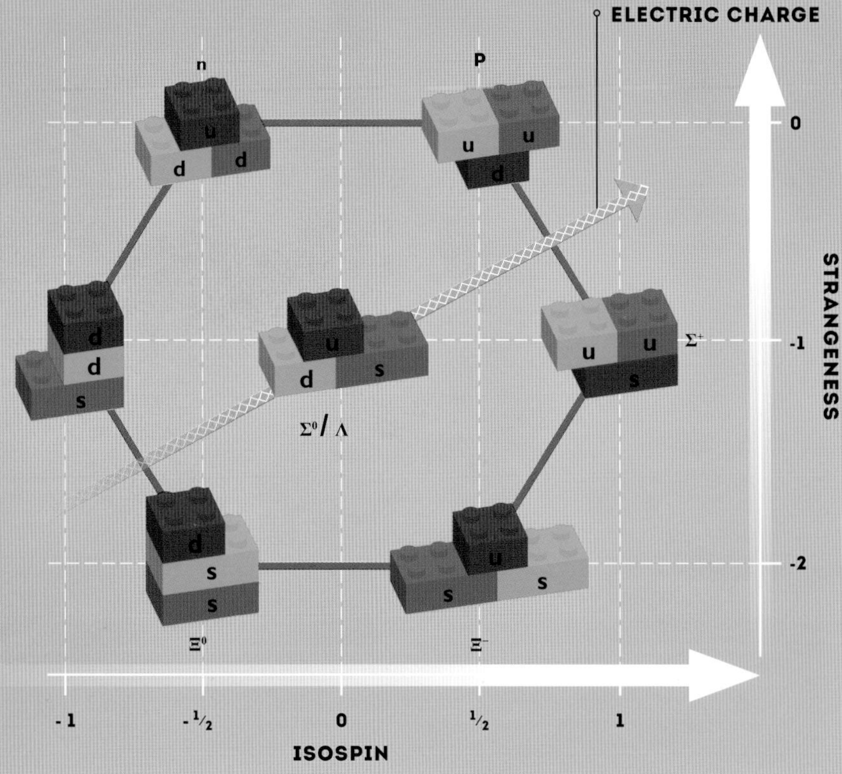

The group of ten baryons were excited, higher energy versions, of the group of eight. In these particles the quantum spins of the quarks were all facing in the same direction, giving them an overall spin of $1/2 + 1/2 + 1/2 = 3/2$. If they are all aligned, then this also meant that new baryons made from three of the same quarks can exist.

The nucleons (neutron and proton) made from only up and down quarks have excited versions in the form of the delta Δ particles. Particles heavier than these nucleons, the sigma Σ and xi Ξ hyperons, have excited versions denoted by the addition of an asterisk. The predicted new omega minus, Ω^-, hyperon has a unique name.

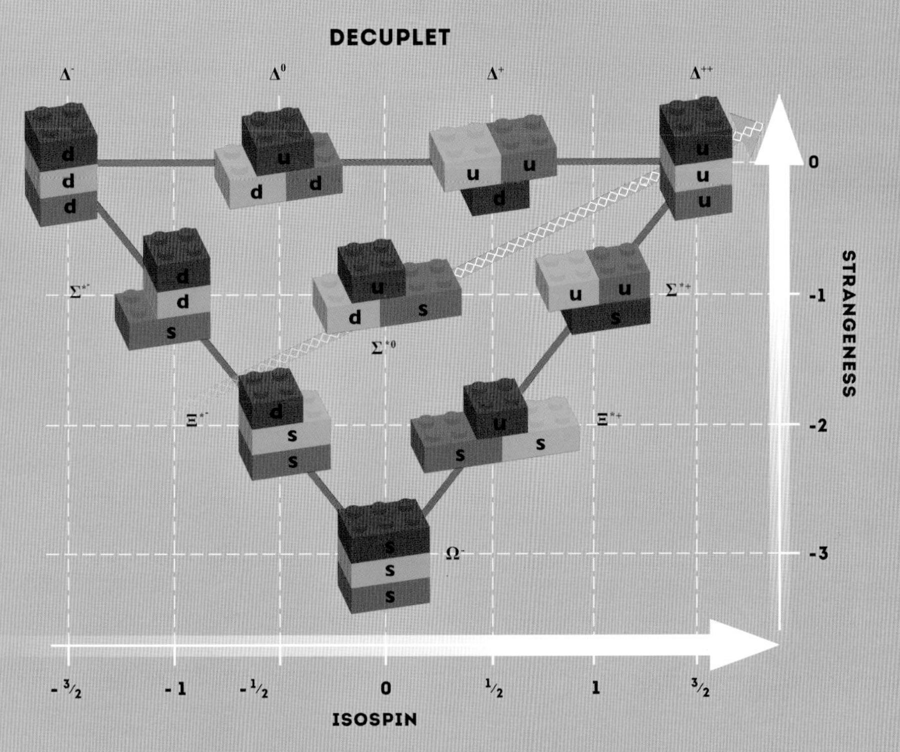

OMEGA MINUS

This was a fantastic theory because it was predictive. It predicted the existence of a triple strange particle with negative charge – the omega minus, Ω^-. Immediately after Gell-Mann presented the idea, the search for the omega began, and it wasn't long until it was seen decaying into a xi-zero (Ξ^0) and a negative pion (π^-).

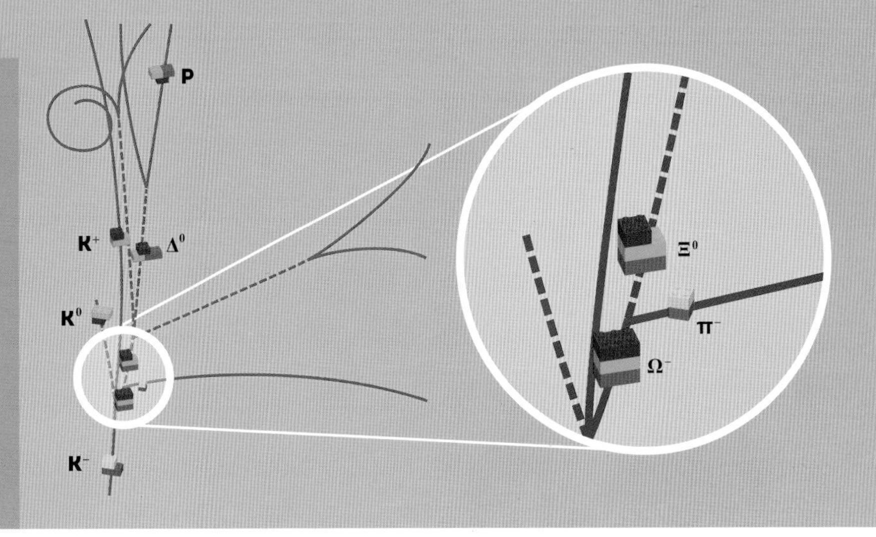

Symmetry between particles and antiparticles led to predictions of opposites for all particles

ANTIBARYONS

The discovery of the antineutron and antiproton demonstrated that all particles, not just the leptons, have an antimatter version. It is a logical extension then to assume that there exist an octet of spin ½ and a decuplet of excited spin 3/2 antibaryons. Comparison of how hyperons and antihyperons are produced and then decay may provide a path beyond the Standard Model. Any difference in the behaviour of the particles and antiparticles suggests some new physics.

OCTET

As before, antiparticles are denoted by a bar above the particle symbol. Note that both the isospin and the strangeness have reversed, as one would expect because the antiparticles are mirror opposites to particles.

There are two unique particles in the centre of the octet, despite being made from the same combination of quarks. Following my highway car analogy on page 85 the Λ^0 is like a car which has zero velocity north–south. It has zero isospin magnitude. The Σ^0 on the other hand is a moving car, with an isospin magnitude of +1, but its speed is being matched. So, although it has an isospin magnitude, it seems to have an isospin magnitude and direction of zero. This why for the same quark content we observe two totally separate particles.

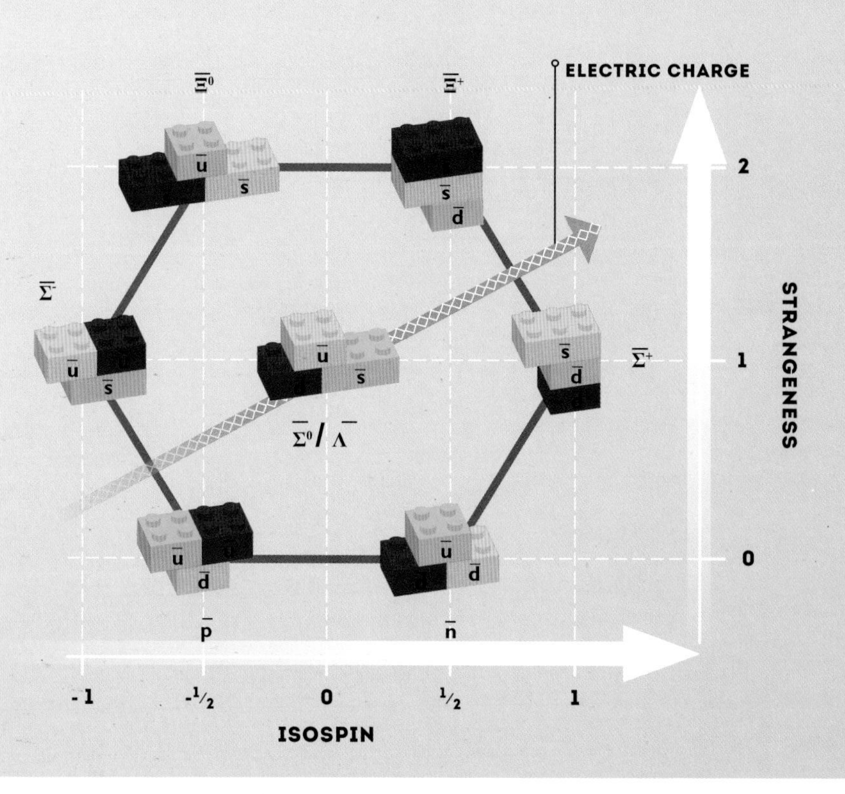

DECUPLET

The antibaryon decuplet contains all of the excited spin 3/2 particles made from three antiquarks. Each antibaryon has an opposite charge to its mirrored baryon version because the quarks from which they are made have entirely opposite electric charges. Antiup quarks have a charge of -2/3 while down and strange quarks have a charge of +1/3.

SPIN

Spin is a property of all particles. It is a totally abstract property only historically linked to the spin of solid objects. All fermion matter-building particles have spin ½ while all boson particles have integer number spins of 0 (Higgs), 1 (photon, gluon, Z and W) and possibly the graviton with spin 2.

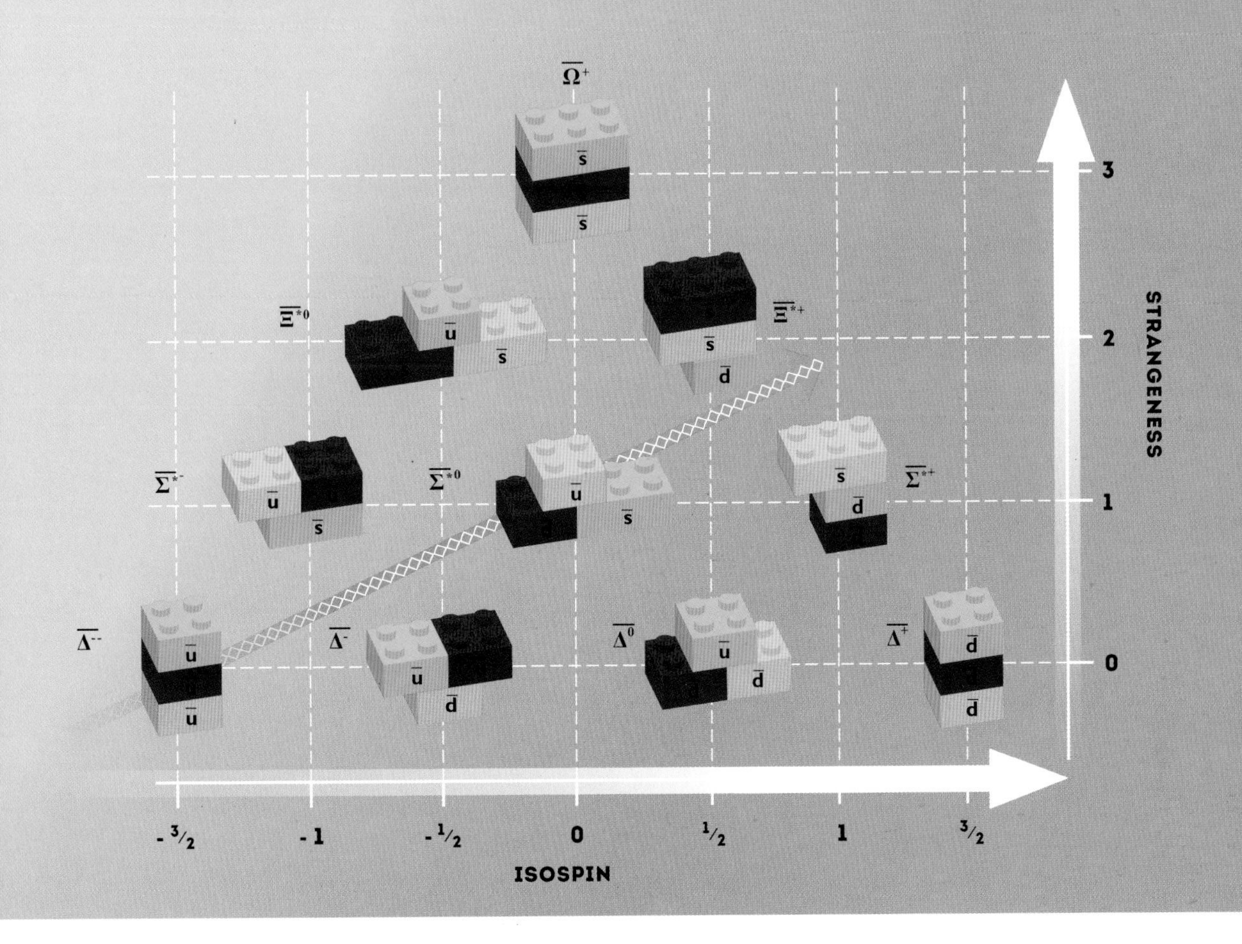

Mesons also fit into the patterns and new excited versions were predicted and discovered

MESONS IN THE EIGHTFOLD WAY

Mesons also fit into Gell-Mann's eightfold way, forming similar patterns to baryons. In the centre there are three possible combinations of up, down and strange and their antiparticle partners which all have values of zero for isospin, strangeness and electric charge. With no strong or electromagnetic identity, they cannot be distinguished by these forces.

Due to the bizarre quantum world, up, down and strange particles and antiparticles mix in different ways to create totally separate observable particles. While the underlying quarks are paired, up–antiup, down–antidown and strange–antistrange, the particles that can be physically measured are combinations of each.

The almost identical mass up–antiup and down–antidown are mixed to form the neutral pion, π^0. This π^0 particle can have the strange–antistrange configuration combined with it in a positive or negative way. These two differences result in two unique particles each with a different mass; the eta (η) meson with a mass of 548 MeV and the heavier eta prime (η') meson with mass 958 MeV.

Understanding the creation and decay of mesons allows us a deeper insight into how the strong force works. The mixing of the different types of meson also provides a direct route to probing the imbalance between particles and antiparticles, an essential part of our Universe's creation story.

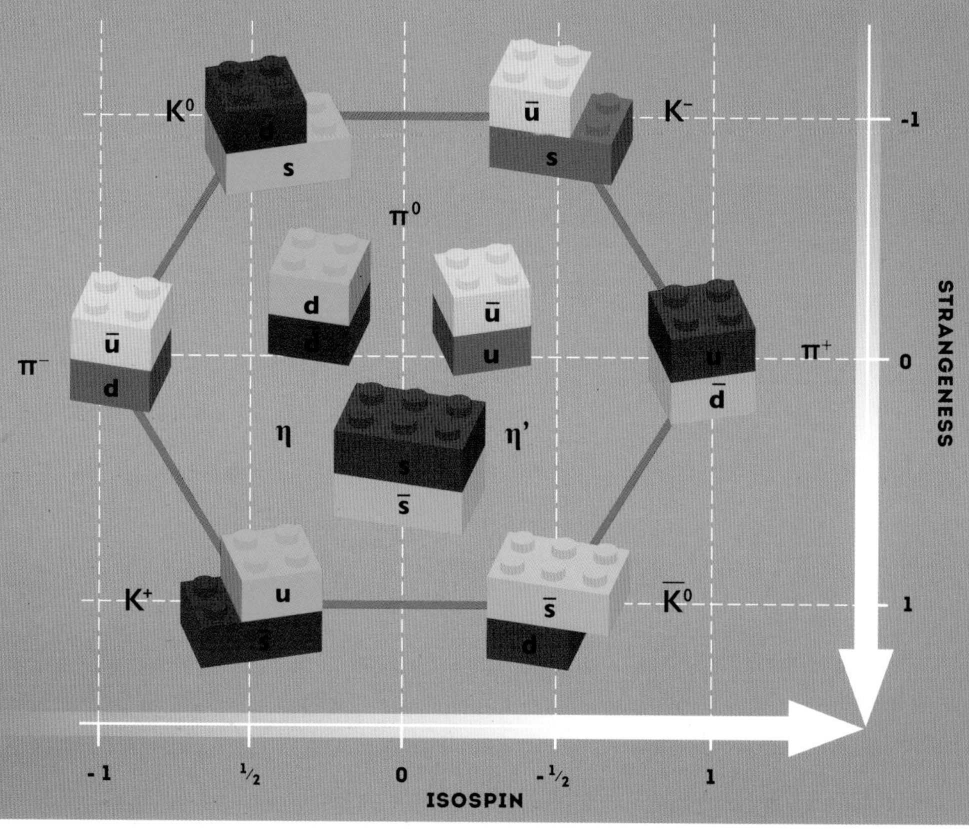

EXCITED RESONANCE

Just like baryons, mesons can also be found in excited resonance versions. The excited pions are called rho mesons, ρ. The excited version of the eta mesons are the omega ω and psi ψ mesons. Excited versions of the kaon are just given a star – K^+ becomes K^{*+}. Baryon and meson excited versions are different from the unexcited versions because of the alignment of the spin of the quarks from which they are made.

Try and line up two or three magnets, all with their north pole pointing in the same direction, and you will feel the tension as they try to change their alignment. This tense state is high in energy, requiring you to hold them there, and if you let go one magnet will flip around and there will no longer be tension. They'll be stable.

The same thing occurs with quarks within baryons and mesons. If the spins of all the quarks are aligned, then they are in a higher energy and tense state. These are the excited resonances and the higher-energy tense state leads to their higher masses. After all $E=mc^2$.

MESONS

Mesons, as predicted by Yukawa, would have a mass between that of a proton and an electron. Although this is true for pions, not all mesons fit this description, and so the modern definition of a meson is very different to Yukawa's. All particles, no matter their mass, are a meson if they contain one quark and one antiquark.

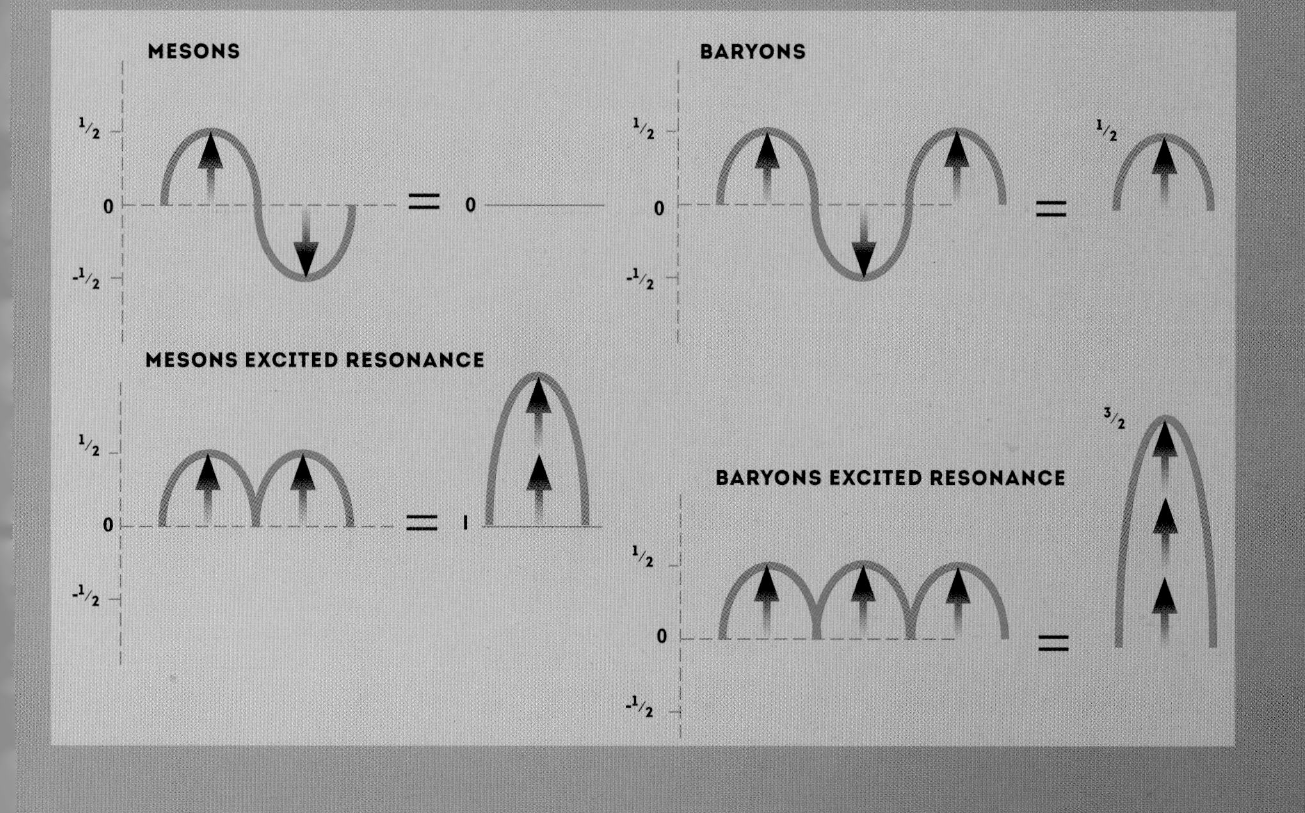

More powerful particle accelerators uncovered a new, heavier, type of quark

CHARM QUARKS

Soon new particles were being discovered which didn't fit this pattern. The first was discovered almost simultaneously by two US groups at Brookhaven National Lab in New York fronted by Sam Ting and at the Stanford Linear Accelerator Centre in California headed by Burton Richter. Following Greek lettering convention, Richter called the particle psi Ψ while Ting called it the J, which was very similar in shape to the Chinese character for Ting's name (丁). Still the only particle to have a double name, it is called the J/Ψ.

The charm is the partner to the strange quark as the up quark is the partner to the down. We can extend Gell-Mann's model to include the charm quark, adding a new dimension in a new conserved number, analogous to strangeness, called charmness. This additional symmetry creates more complicated patterns of baryons and mesons.

By this point isospin had been predicted because of the existence of the neutron, strangeness because of the kaon, and charmness because of the J/Ψ. All of these particles were simply placed on a pattern represented by relevant symmetries, but the underlying structure was not understood. However, it was soon realized, by Gell-Mann again, that these conserved quantities of isospin, strangeness and now charmness were related to smaller structures inside the particles which he named quarks.

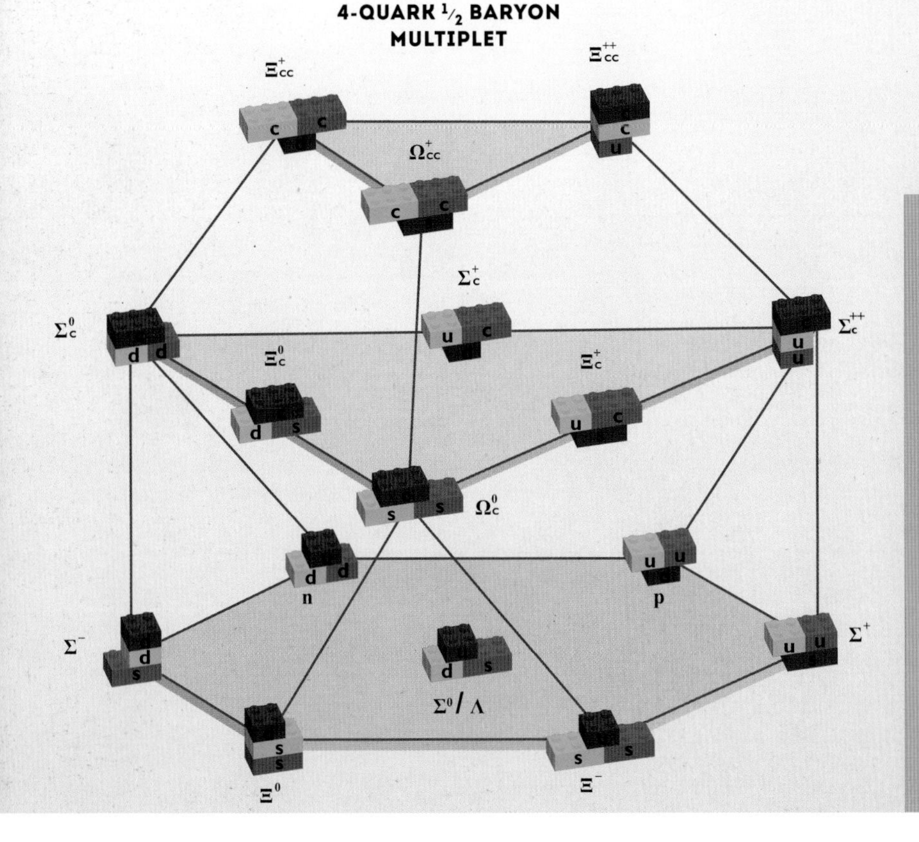

THREE QUARKS FOR MUSTER MARK!

Murray Gell-Mann had the sound he wanted for the name of the particles which made up baryons, but not the written word, until he read James Joyce's *Finnegans Wake*. One passage read "three quarks for Muster Mark!", the word quark fit Gell-Mann's sound and the mention of three was happy coincidence with the number of quarks in a baryon. Joyce's quark is a raucous cheer, Gell-Mann's quark is a fundamental particle.

With all these different quarks and their conserved numbers, the shapes created would be very complicated. Luckily all of these patterns, however complicated, were formulated into a theory which explained this zoo of particles in just a few simple rules.

4-QUARK $^3/_2$ BARYON MULTIPLET

4-QUARK MESON MULTIPLET

NAMING BARYONS

The naming convention for the new charm-containing baryons was to use the name from the original patterns and just add a subscript c for every charm quark within the baryon.

The strong force has three types of charge, so we use the analogy of primary colours

QUARKS AND COLOUR

All this pattern finding suggested an underlying symmetry. Unlike the electromagnetic force arising from a one-dimensional symmetry with just positive or negative charge, the symmetry describing the strong force was three-dimensional. The same as there are three directions in space, there had to be three types of charge associated with the strong force. Every quark interacted with the strong force and so carried this strong charge.

Searching for an analogy, physicists struck upon the combining of primary colours of light. Each of the three strong charges were assigned a primary colour; red, green and blue. This colour theory of the strong force is called quantum chromodynamics or QCD for short; quantum for the behaviour of the particles, chromo as it is based on colour and dynamics as it dictates how the particles move.

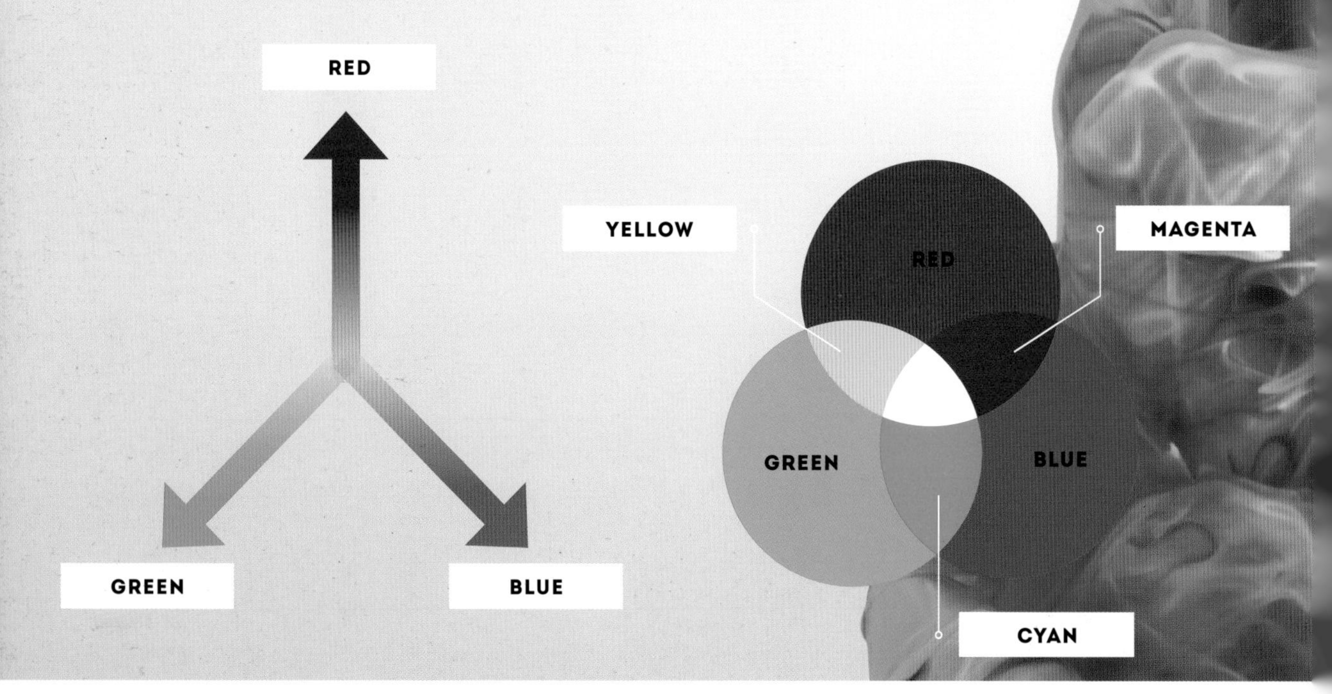

STRONG FORCE

The strong force binds quarks together into hadron particles but also binds protons and neutrons together in atomic nuclei. It is named strong because it has the strongest interactions of any force.

The same rule applied for these charges as for electric charges – like charges repel each other while different charges attract. This resulted in three quarks, each with different a different colour charge, attracting each other.

DIFFERENT COLOURED CHARGES ATTRACT

LIKE COLOURED CHARGES REPEL

Combining the three primary colours of light results in colour-neutral white light. Combining the three colour charges of quark results in a colour-neutral particle. Once neutral, the same way that an electrically neutral atom does not attract any further electrons, these particles do not attract any further colour-charged quarks toward it. All particles made from quarks (hadrons) are colour-neutral.

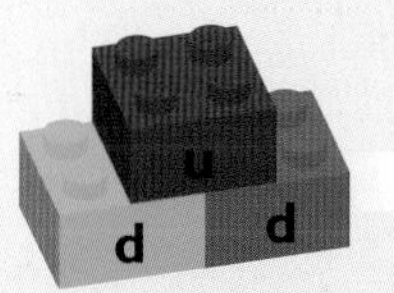

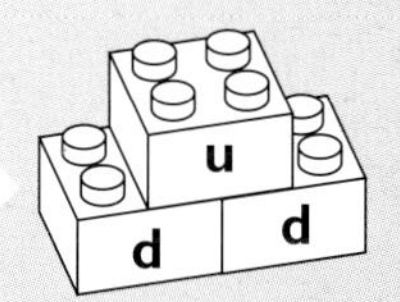

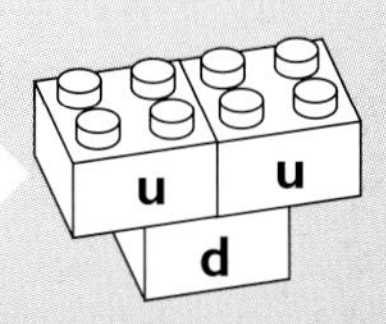

This is the reason that all baryons are made from just three quarks, directly echoing the three-dimensional symmetry of the strong force – one red, one green and one blue.

Anticolour charges are opposite in every way to a colour charge

ANTIQUARKS AND ANTICOLOUR

The negative or anticolour charges can be represented by the secondary colours of light: cyan, magenta and yellow.

Cyan is a combination of green and blue light, magenta a combination of red and blue, and yellow an equal mix of red and green.

ANTIMATTER

Antimatter is the complete mirror opposite version of matter, interacting with the forces of Nature in totally the opposite way. Every particle has an antiparticle opposite number. Antiquarks therefore have anticolour, the opposite of colour charge.

YELLOW

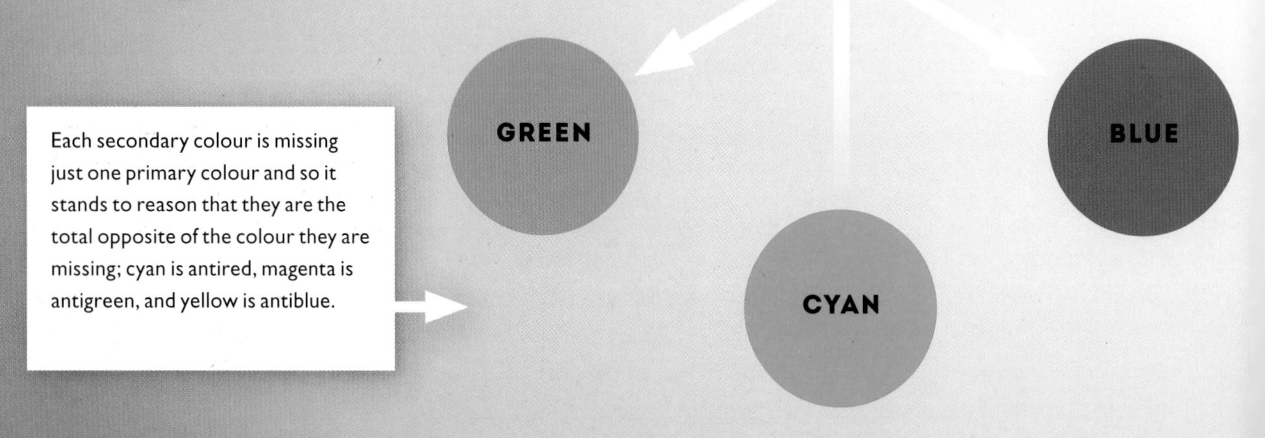

Each secondary colour is missing just one primary colour and so it stands to reason that they are the total opposite of the colour they are missing; cyan is antired, magenta is antigreen, and yellow is antiblue.

Quark–antiquark pairings make up meson particles. These combinations result from the same symmetry as the baryons which is the reason they are arranged in a very similar way. Baryons and mesons are all made of quarks and any particle made from quarks is also given the name of hadrons – this is the H in the LHC which collides protons.

Combining a red quark with a cyan (antired) antiquark produces a colour-neutral, white-light, particle.

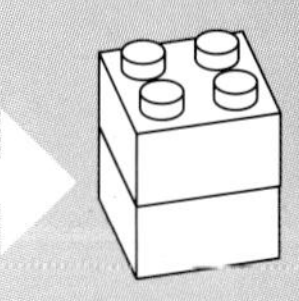

Combining a blue quark with a yellow (antiblue) antiquark has the same effect.

In the same way, combining a green quark with a magenta (antigreen) antiquark produces a white-light particle.

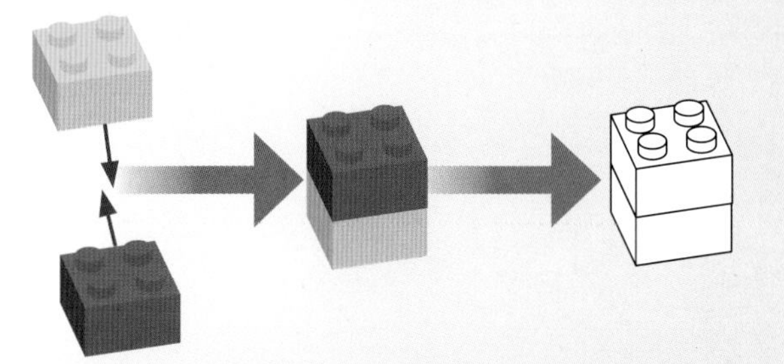

All baryons have antiparticle versions of themselves made from three anticolour antiquarks.

Colourless, white particles contain equal amounts of the three primary colours of light to create white light. The only way to achieve this with the secondary anticolours is to combine all three to create antibaryons.

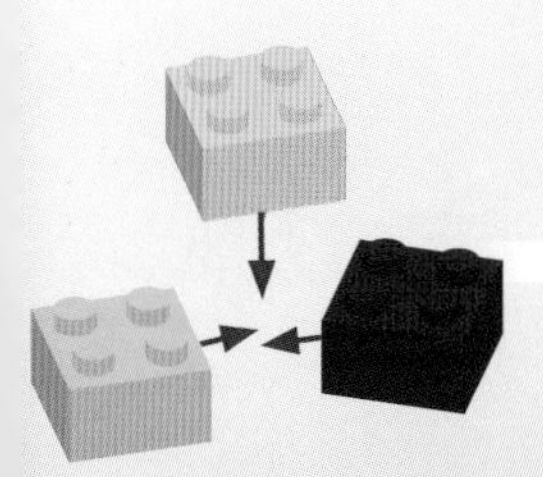

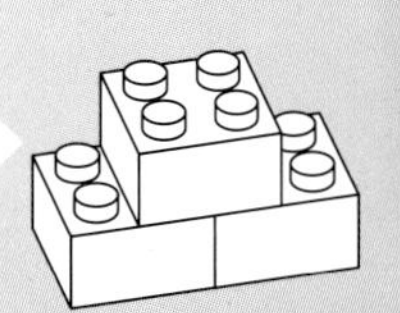

Gluons transfer energy, strong colour and anticolour charges

GLUONS AND STRONG CHARGE EXCHANGE

In QCD the strong force is exchanged between quarks by bosons called gluons. Unlike the photon, which exchanges the electromagnetic force, gluons carry charge – strong colour charge. Gluons are combinations of colour and anticolour. This means that an exchange of a gluon between quarks not only attracts quarks to one another but also changes their colour charge.

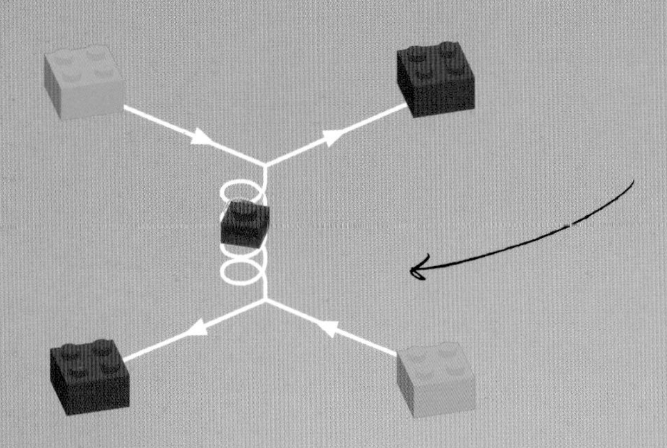

Just like photons, gluons have no mass. Photons carry the electromagnetic force but cannot interact with each other as they do not carry electric charge. Gluons, on the other hand, do carry colour charge and so can interact with themselves. To distinguish their different behaviour from photons, gluons are represented on Feynman diagrams by curly lines. Gluons can interact with gluons in the following Feynman diagrams.

Take away negative electric charge from an atom and you leave it more electrically positive than it was before. If you remove a negative strong anticolour charge from a quark then the quark must take up the opposite strong colour charge. To change a red quark to green you need a gluon to take away the red charge but also negative antigreen (magenta) to leave it greener than before (see page 110 to prove this is the case).

If this red–antigreen gluon then interacts with a green quark the antigreen cancels out its greenness and the red in the gluon colours the quark red. We started with one red and one green quark, we have ended with one red and one green quark but the gluon has been exchanged. In this way the strong force is being exchanged constantly, attracting quarks together into stable particles.

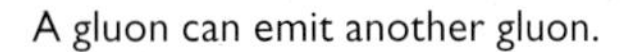

A gluon can emit another gluon.

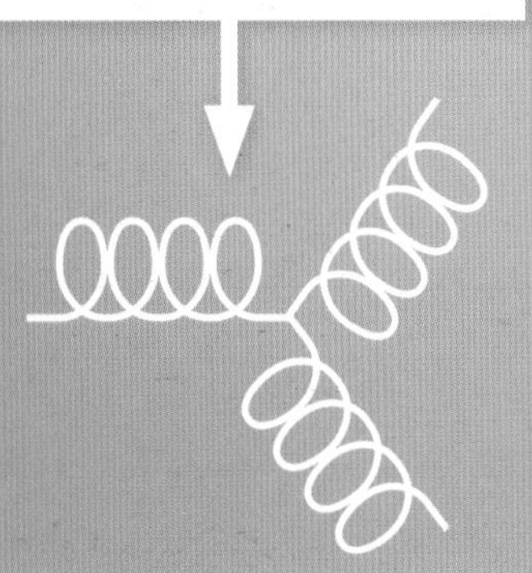

Two gluons may combine to form a single gluon.

Two gluons may scatter off of one another.

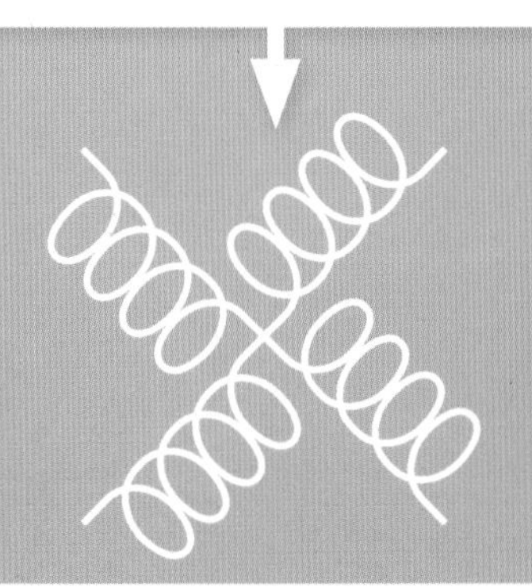

The following are some of the ways that a gluon can interact with quarks in Feynman diagrams.

Note that the gluon is a combination of the colour of the incoming quark and the anticolour of the outgoing quark, this is the only way that colour is conserved at the vertex at which the gluon interacts with the quark. And of course we can rotate the quark lines to get the analogous diagrams for antiquarks.

FEYNMAN DIAGRAMS

Feynman diagrams are technical cartoons which sketch out possible interactions between particles. Time passes along the horizontal axis and the vertical axis represents space.

The yellow colour charge of the antiquark is taken away by the gluon. The gluon also takes away antimagenta (green) which makes it more magenta than before.

The red colour charge of the quark is taken away by the gluon. The gluon also takes away antiblue (yellow) which makes it bluer than before.

The original red colour charge of the quark is taken away by the gluon. The gluon also takes away antigreen (magenta) which makes it greener than before.

The cyan colour charge of the antiquark is taken away by the gluon. The gluon also takes away antiyellow (blue) which makes it more yellow than before.

GLUEBALLS

The self-interaction of gluons leads to the prediction of a new form of exotic matter called a glueball. Glueballs are made entirely from gluons and so are different to quarks because they are electrically neutral. While glueballs have not yet been observed, their existence is suggested in data from a number of different experiments. The search continues today at the LHC and other dedicated experiments where the focus is on the possible observation of glueballs decaying into mesons.

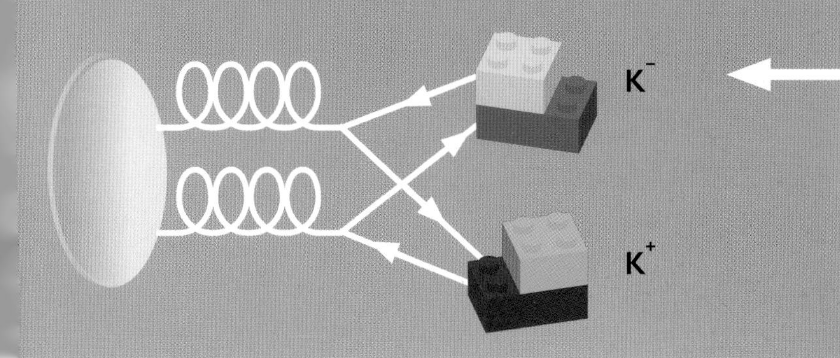

Gluons, like any other boson, can produce particle–antiparticle pairs. Because gluons interact only with quarks they produce quark–antiquark pairs. Gluons from a glueball would produce numerous pairs of quark–antiquark, forming mesons which could then be detected.

As quarks are pulled apart, they produce quark-antiquark pairs

CONFINEMENT

When we talked about the electromagnetic potential energy hill between two electric charges, we found that the hill flattened out as the charges were pulled apart (see page 65). This is a direct effect of the electromagnetic force becoming weaker at greater distances as its effect is spread out – it actually reduces as 1 divided by the distance squared. The strong force, however, acts with a constant attractive pull directly along a gluon line connecting two quarks. It does not become weaker with distance.

Pulling apart quarks attached by the exchange of gluons, raises them in energy as they climb up a continually increasing potential hill.

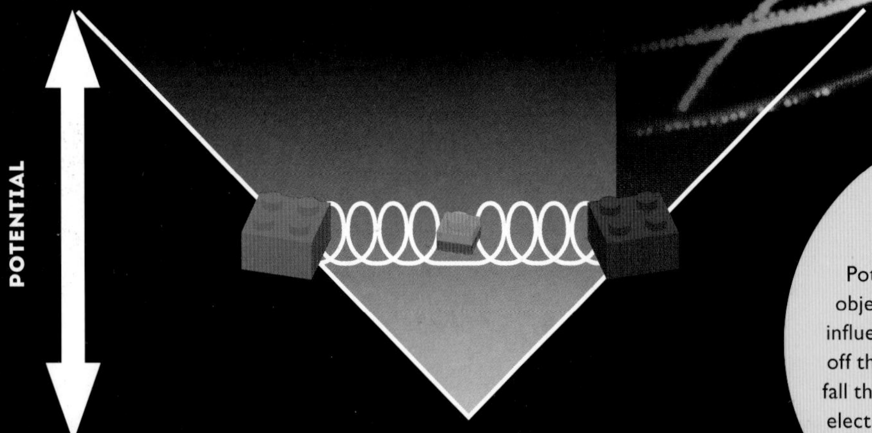

POTENTIAL

POTENTIAL ENERGY

Potential energy is the energy an object can transfer if allowed to be influenced by a force. Raise an object off the ground and it has potential to fall thanks to the force of gravity. Two electric charges have the potential to attract or repel one another.

This means that you would need more and more energy to pull quarks further and further apart. The constant force means that the potential energy hill of the strong force continues to rise as quarks are separated and the attractive gluon between them is stretched. Very quickly the potential energy becomes so large that it is possible to create a new quark–antiquark pair. These new particles will be close to each other and therefore have a lower potential energy, effectively resetting the potential energy hill. The distance at which this happens is tiny, much smaller than the size of a proton, and so quarks can never be stretched out of a hadron to be observed separately and alone.

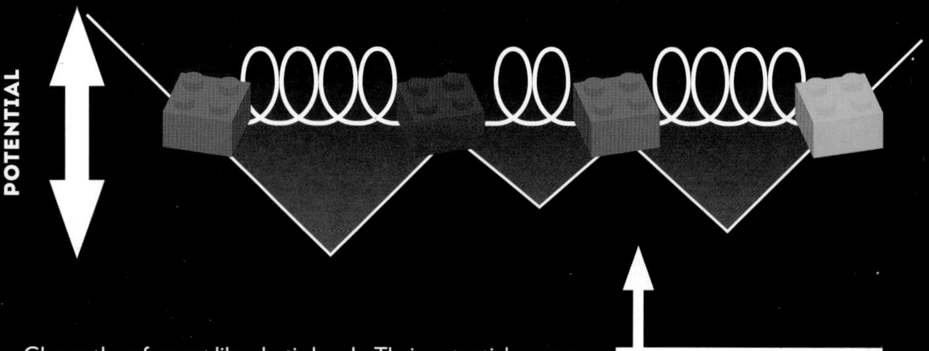

POTENTIAL

Gluons therefore act like elastic bands. Their potential energy increases until a breaking point when they snap to form a quark–antiquark pair. This continues until the gluon has given up all its energy into the formation of quarks and antiquarks. This creation of quark–antiquark pairs and the strength of the attractive colour charges they have does not allow any quark to escape from another. They are instead confined into groups to form colourless mesons or baryons.

Soon the quarks have climbed in energy up the potential hill high enough that they have enough energy to create the mass of a quark–antiquark pair.

The image on the left is a reconstruction of an event in the DELPHI detector. It clearly shows three jets of hadron particles formed from energetic quarks trying to escape the original particles they were in. Along the jets pairs of quark and antiquark create cascades of hadron particles

JETS

The distance at which gluons snap is far smaller than the size of a proton. When we try to knock quarks or gluons from inside of a baryon by colliding protons together, we instead form long chains of quark–antiquark pairs which themselves form new baryons and mesons. These chains of new hadron particles are called jets. Their direction and energy allow physicists to determine the energy and direction of the original quark or gluon responsible.

Most of the mass of protons and neutrons comes from gluons and quark–antiquark pairs

RESIDUAL STRONG FORCE

You may or may not have noticed a large difference between the mass of up and down quarks, and the protons and neutrons. Protons with mass of 938 MeV and neutrons with mass about 940 MeV are much heavier than simply a combination of just three up and down quarks. Up quarks with a mass around 2.3 MeV and down quarks with mass around 4.8 MeV would only contribute a small percentage of the overall mass of protons and neutrons. Where is all of the extra mass coming from?

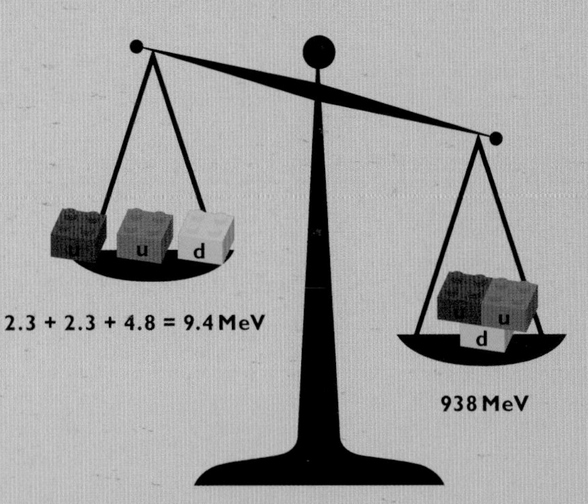

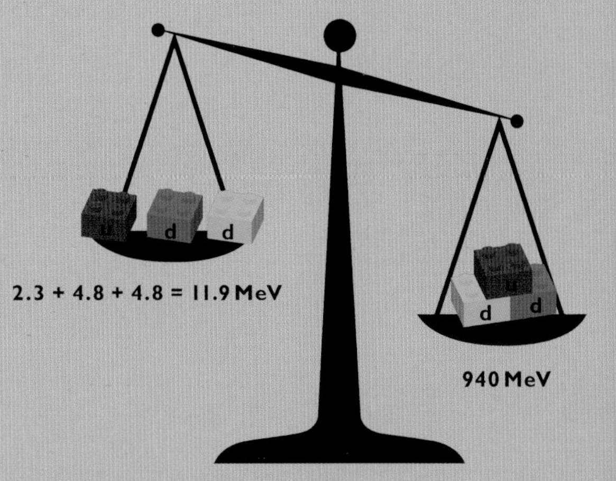

UNCERTAINTY PRINCIPLE

Almost the entire mass of a proton or neutron is down to a quantum effect known as the uncertainty principle which effectively tells us that our Universe is pixellated. The same way plastic bricks are limited to be multiples of at least one stud, pairs of properties of particles are limited too. When observing a particle you cannot measure both the position and momentum of the particle with perfect accuracy. The combined uncertainty in these measurements is given by a number called Planck's constant ($\hbar$), named after quantum physics' founding father, Max Planck. It defines the finest resolution possible of pairs of measurable quantities; position and momentum or energy and time. The uncertainty principle restricting the location of the up and down quarks leads to them having extra energy. This extra energy then manifests itself as the exchange of lots of gluons.

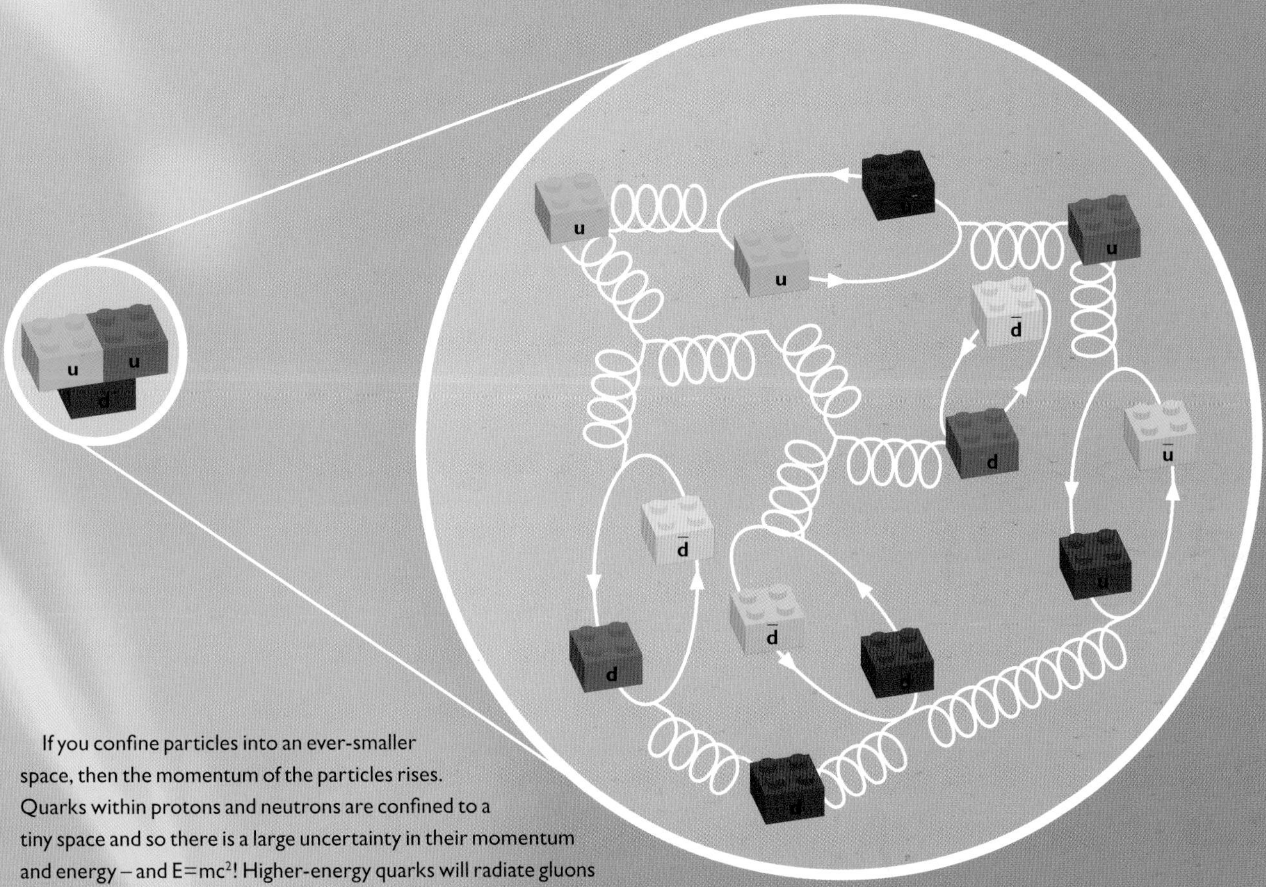

If you confine particles into an ever-smaller space, then the momentum of the particles rises. Quarks within protons and neutrons are confined to a tiny space and so there is a large uncertainty in their momentum and energy – and $E=mc^2$! Higher-energy quarks will radiate gluons which will quickly split into quark–antiquark pairs. It is this sea of gluons, quarks and antiquarks, constantly bubbling around the three main quarks, which accounts for over 95% of the mass of a proton or neutrons.

LATTICE QCD

The combination of the strong force being so strong and the self interaction of quarks means that calculation of strong force particle interactions becomes very complex very quickly. The best current method for determining such strong interactions is something known as lattice QCD. Quarks are modelled as points on a lattice, like the crossovers on squared paper, while gluons are modelled at the linking lines between. Calculations of strong force interactions like this require many computing hours, increasing just less than exponentially as the detail of the calculations increase. Despite this, lattice QCD has had many successes, including successfully modelling the binding energy of the quarks in a proton to get its mass to within 2% error.

PIXELATION OF NATURE

This pixelation of Nature represents the horizon at which the Standard Model and all quantum theories can operate. It means that things in our Universe where there is very high energy in a very small space, such as is found in black holes or at the Big Bang, cannot be explained successfully using this theory. This places a real upper limit to the highest energy and smallest scales at which the Standard Model works – values known as the Planck energy and the Planck length. Things with energy higher than 1.22×10^{19}GeV, or taking place in spaces smaller than 1.62×10^{-35}m fall outside of the explanation of the Standard Model.

Tetraquarks and pentaquarks can also be formed from combinations of quarks

TETRAQUARKS AND PENTAQUARKS

Combining quarks with the three primary (or secondary) colours to make baryons or combining one colour and anticolour to make mesons are the two most basic ways of making colour-neutral particles. But why should it stop there? Why not continue piecing together ever more quarks, as protons are combined with electrons to create ever heavier neutral atoms. The sea of gluons which binds quarks together gives such exotic new particles very large masses, which makes them very unstable. It is only in the last few decades that evidence of these exotic particles has been observed.

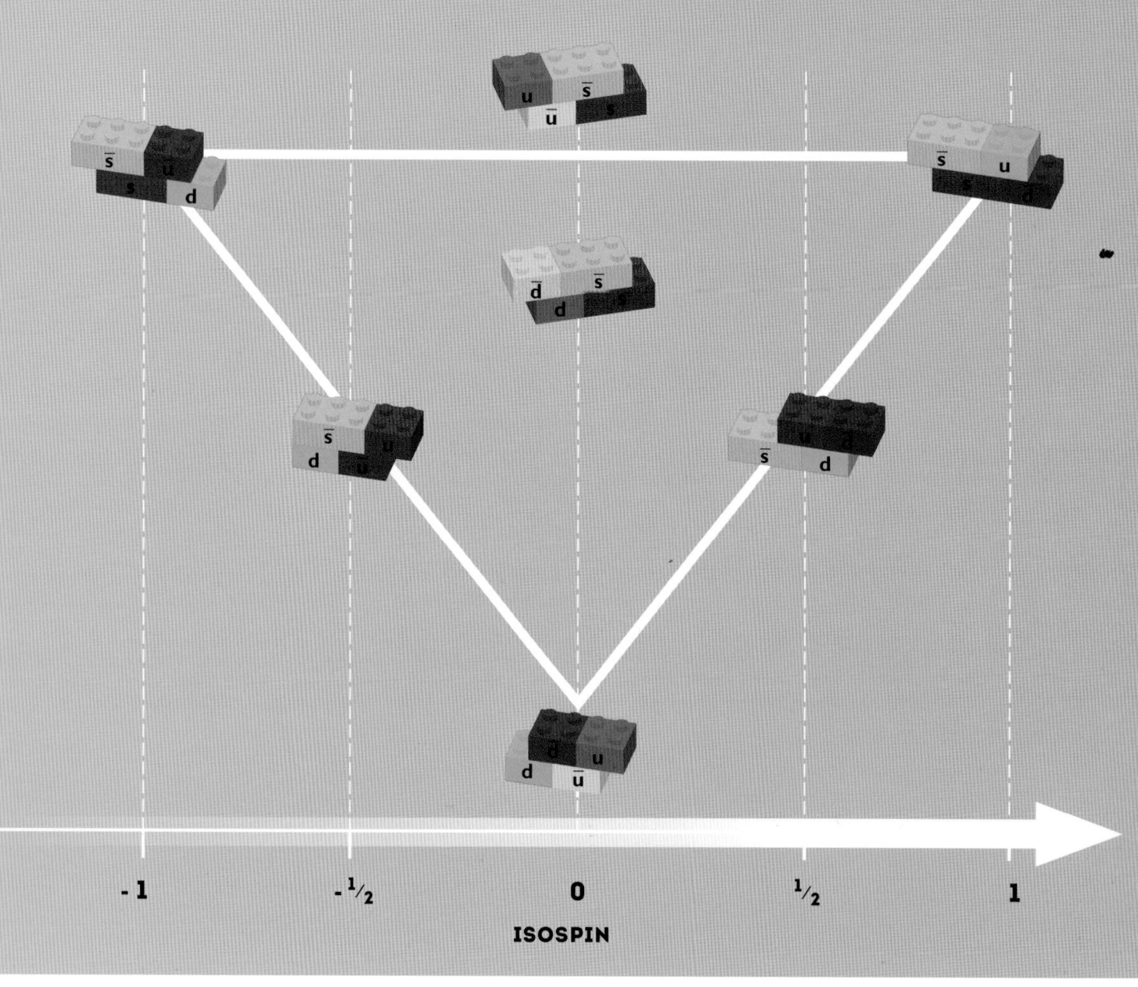

TETRAQUARKS

Mesons combine one quark and one antiquark of colour and anticolour to create a colourless particle. Adding one more quark and one antiquark results in a group of four particles which are still colour-neutral. These exotic meson particle states are called tetraquarks, and they were first seen in 2013 at the BES III experiment in China and the Belle experiment in Japan. Since then the LHCb experiment at the Large Hadron Collider has confirmed the discovery and added more tetraquarks to the list. On the left is a tetraquark version of Gell-Mann's eightfold way.

PENTAQUARKS

Another possible configuration comes from adding a quark and an antiquark to the three quarks of a colour-neutral baryon. These five quark states are called pentaquarks which were suggested soon after the development of the quark model. There were claims early this century of a pentaquark, the theta plus, Θ^{+}, but this could not be confirmed. In 2015, however, the LHCb experiment showed without reasonable doubt the existence of a different pentaquark, which was called the P (for… you guessed it pentaquark) – P_c.

It is not known if these quarks form a single particle, like the proton, or if it is a fusion of a baryon and meson into a baryon–meson molecule, similar to hydrogen atoms binding to oxygen to form molecules of water. Determining the exact construction of pentaquarks will tell us a great deal about the nature of the strong force, which remains the least understood of the forces of Nature.

Pc – TRUE PENTAQUARK?

4 QUARKS, 1 ANTIQUARK:

2 x up + 1 x down + 1 x charm, and 1 x anticharm;

1 x red + 1 x green + 1 x blue + 1 x red + 1 x antired (cyan)

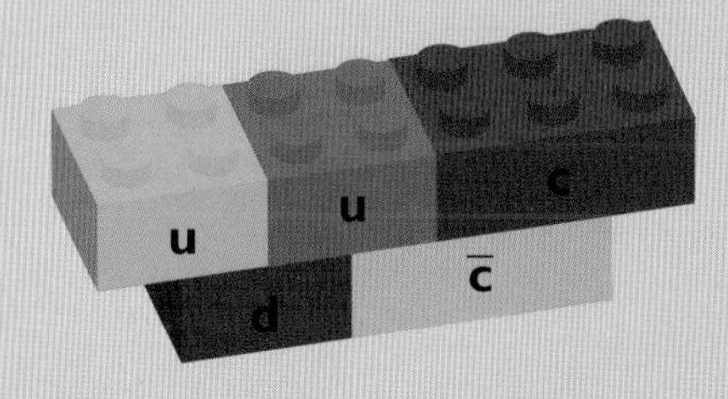

Pc – BARYON-MESON MOLECULE?

1 BARYON (PROTON), 1 MESON (J/Ψ):

2 x up + 1 x down, and 1 x charm + 1 x anticharm;

1 x red + 1 x green + 1 x blue + 1 x red + 1 x antired (cyan)

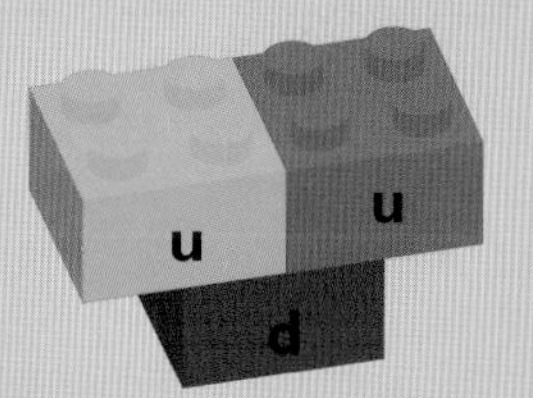

The search is now on for other pentaquarks which can be predicted by following Gell-Mann's patterns.

What is clear, however, is that the study of these bizarre things will give a clearer picture of how the strong force works and therefore how it came to bind the quarks into the seeds of atomic nuclei a tiny fraction of a second after the Big Bang.

QUARKS

Quarks account for one half of the fermion matter-building particles. Quarks experience the strong force and so carry strong colour charge.

STRONG FORCE AND QCD SUMMARY

The strong force determines how a whole array of particles are built and following these rules you can create any possible particle made from quarks.

The strong force is three dimensional in its symmetry and is modelled by the primary colours of light, with antiquarks carrying the secondary anticolours of light.

Opposite electrically charged particles attract while like charged particles repel one another through the electromagnetic force.

All particles made from quarks are called hadrons, all hadrons have to be colour neutral like the primary colours mixing to form white light.

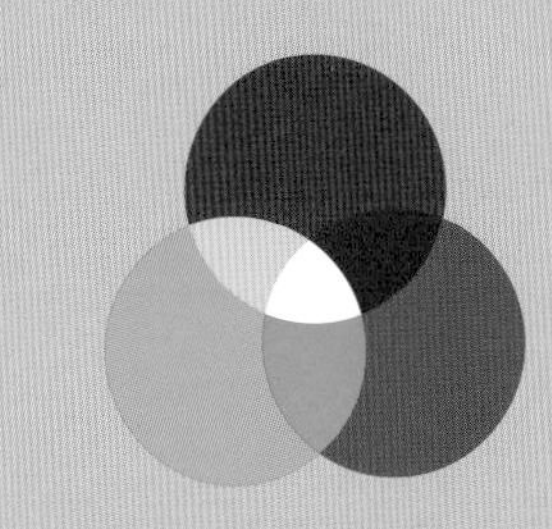

Particles made from three quarks are called baryons. They are made from one red, one green and one blue quark. We cannot tell from experiment which quark carries which colour so all combinations are possible.

Particles made from a quark–antiquark pair are called mesons. Combining a quark with an antiquark means the colours cancel each other out to produce a colour-neutral particle.

Exotic particles also exist. Tetraquarks resemble two mesons bound together, made from two quarks and two antiquarks. Pentaquarks resemble a baryon and a meson bound together, made from four quarks and an antiquark.

All particles which follow the above rules are thought to exist except any containing a top quark. Top quarks have such a short lifetime they decay before even forming a particle.

The strong force is exchanged between quarks by boson particles called gluons, which carry both colour and anticolour.

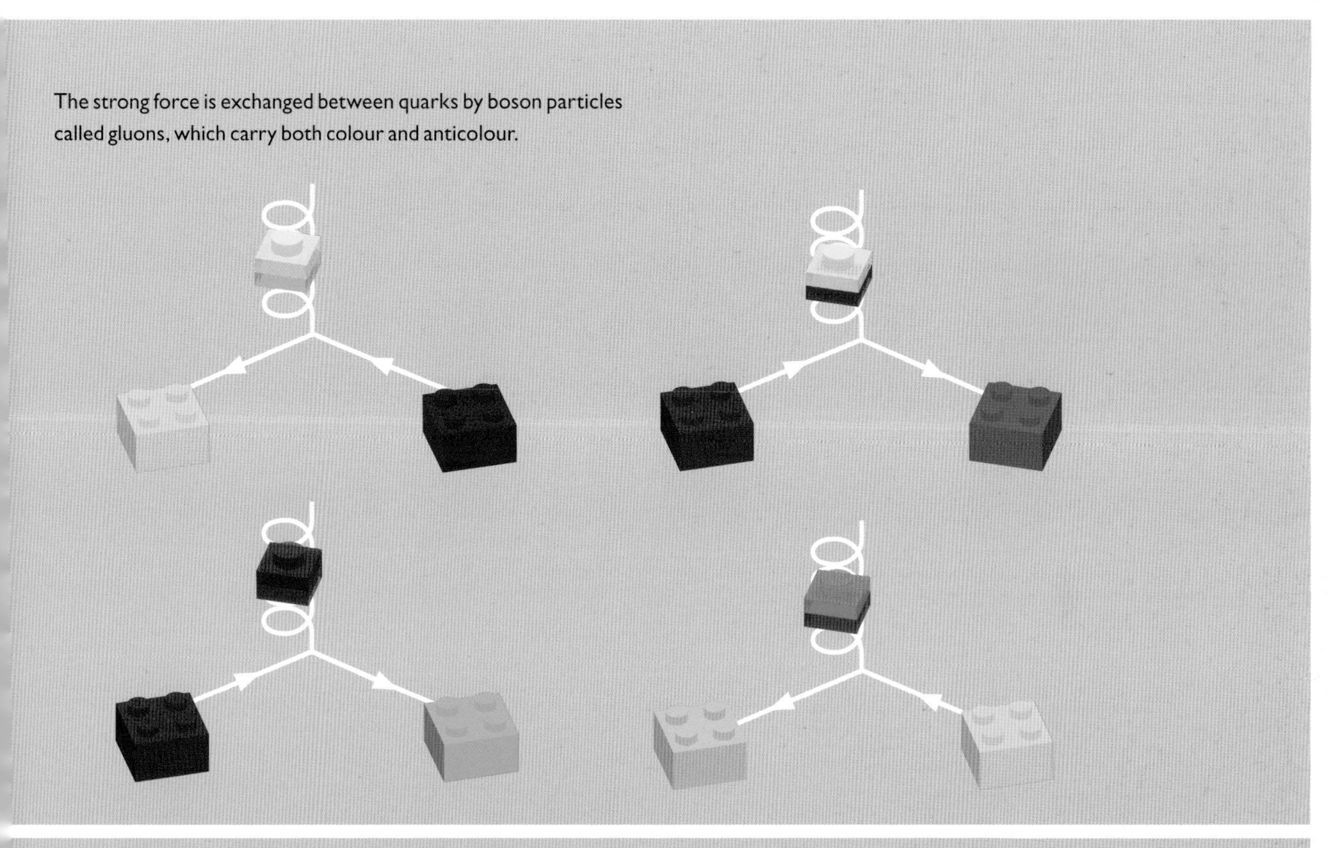

As they carry colour charge, gluons can interact with themselves leading to a predicted exotic form of matter called a glueball which is yet to be observed experimentally.

Gluons continually increase in potential energy as they move away from a colour-charged quark. This results in the creation of quark–antiquark pairs limiting the travel of any gluon to a distance smaller than the width of a proton.

The strong force binding the protons and neutrons together to form atomic nuclei is exchanged by pion meson particles.

Beta decay could not be explained by existing forces and so the weak force was born

BETA DECAY

Beta radioactivity results in a change of chemical element; inside the nucleus a neutron decays into the slightly lighter proton. Zooming into the quarks making up the particles, the change from a neutron to a proton would require one of the quarks to change from a down to an up.

The electromagnetic force does not change a particle. It only exchanges energy between particles with an electric charge.

The strong force does not change quarks from up to down. It only changes the colour charge associated with the quark.

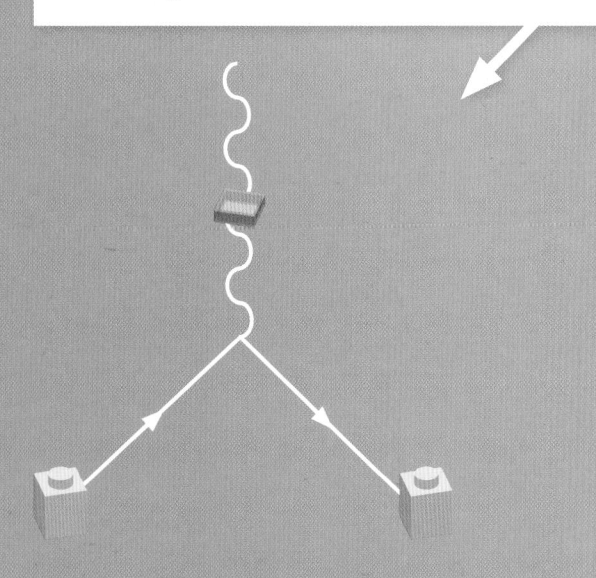

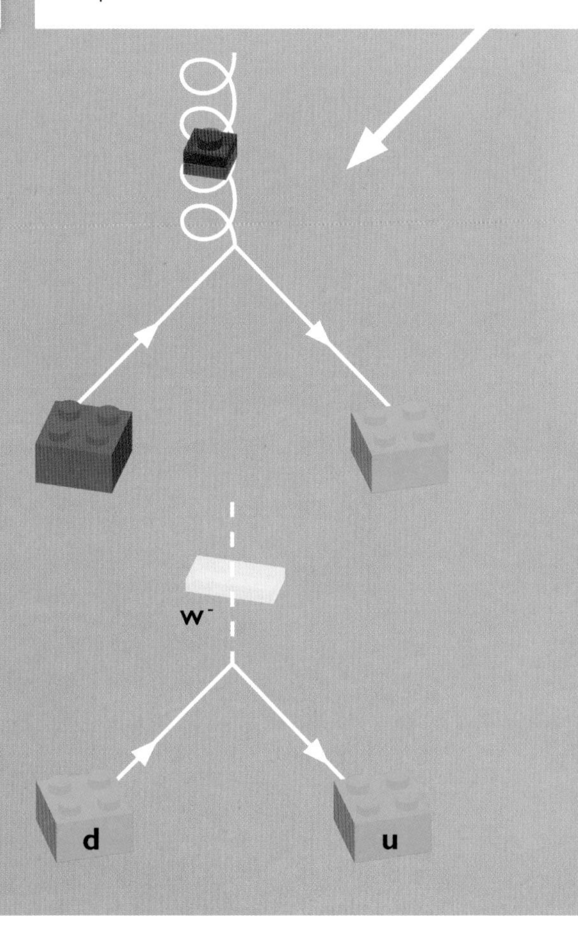

So for such a change to happen we need a new force, the weak force, so called as beta decay happened so rarely it could not be as strong and as likely as a charged particle connecting to photons of electromagnetism. With a new force must come a new boson force-carrying particle – the W, for weak. Changing down quarks, with electric charge of -1/3, to up quarks, with electric charge of +2/3, requires a total change of +1 in electric charge. To balance this change out the new W boson must have an opposite, -1, electric charge.

As well as this change in the nucleus, fast-moving electrons are flung from the atom during beta decay – the beta particles. The energy of these electrons comes from the mass difference of the neutron and proton multiplied by the speed of light squared ($E=mc^2$).

With just electrons emitted in beta decay they should always take away this same energy. Instead, electrons from beta decay were seen to have a whole range of energies from almost zero up to the expected energy. This could only be possible if there was something else taking away a different share of the energy released each time.

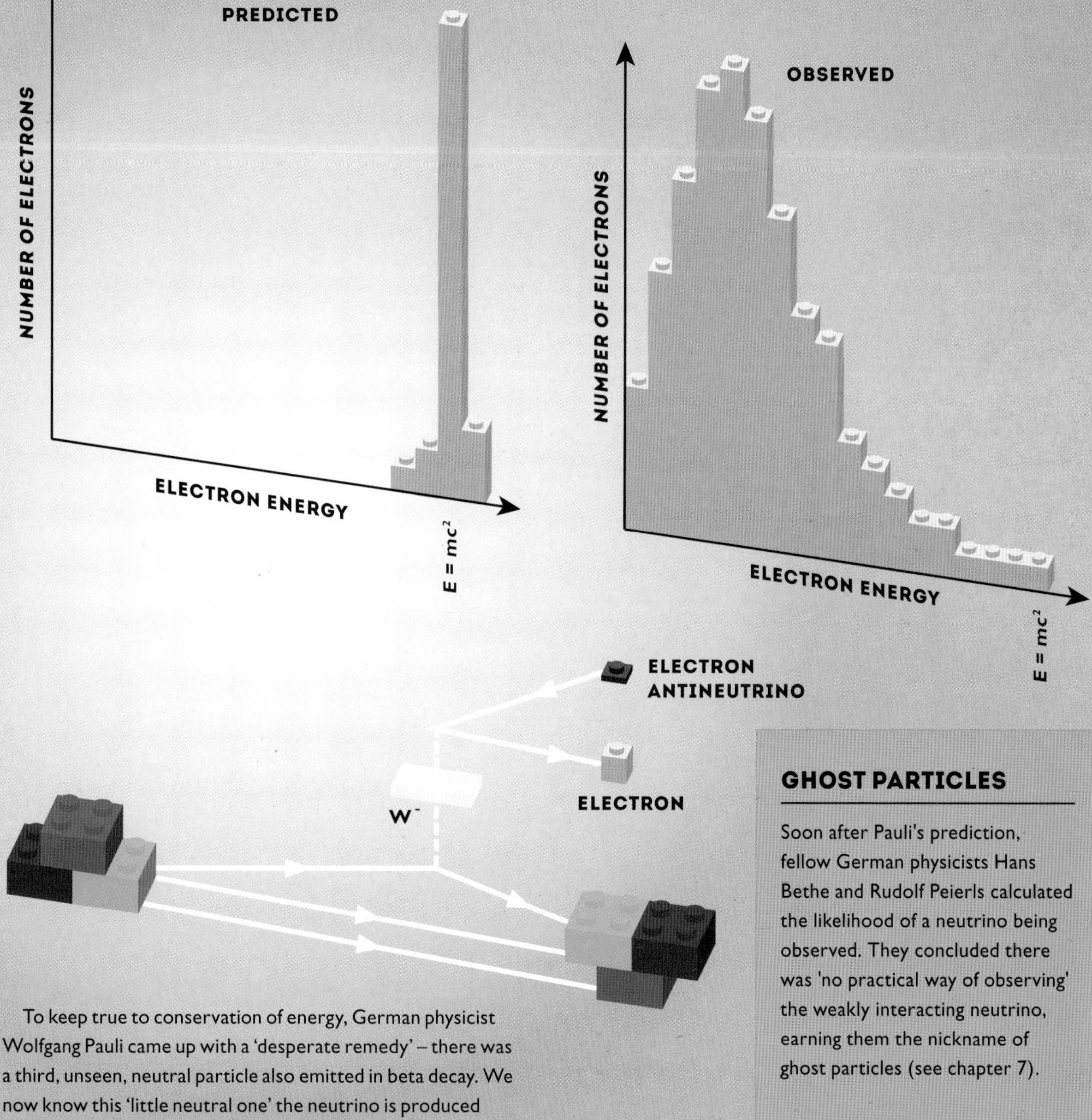

To keep true to conservation of energy, German physicist Wolfgang Pauli came up with a 'desperate remedy' – there was a third, unseen, neutral particle also emitted in beta decay. We now know this 'little neutral one' the neutrino is produced alongside the beta electron by the weak W boson.

GHOST PARTICLES

Soon after Pauli's prediction, fellow German physicists Hans Bethe and Rudolf Peierls calculated the likelihood of a neutrino being observed. They concluded there was 'no practical way of observing' the weakly interacting neutrino, earning them the nickname of ghost particles (see chapter 7).

Weak charge opposites are partner particles

PAIRING UP

Only quarks can feel the strong force, and particles with electric charge feel the electromagnetic force. Neutrinos do not feel either. The presence of quarks, electrons and neutrinos in beta decay shows that all fermion particles must interact with the weak force and W boson.

Looking at beta decay we also start to notice the age-old pattern of pairs, which can only mean one thing; a symmetry and a conserved quantity – pointing us towards the weak force version of charge. While the electromagnetic force is one-dimensional in electric charge and the strong force is three-dimensional in colour, the weak force is two-dimensional like the two sides of a coin.

The weak force groups fermions into oppositely weak charged pairs and the W boson swaps one particle for the other. Beta decay shows us that two of these pairs must be the up and down quark, and the electron and neutrino. Flipping the particles one way requires a W boson with a negative electric charge. So if we change the order of the particles, we find that there has to be a positive W boson that can flip particles the other way.

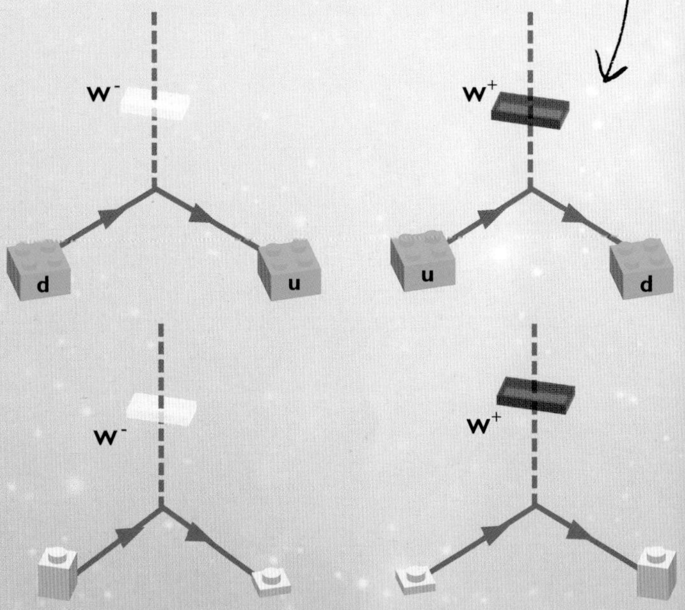

As well as the electrically charged W^- and W^+, which flip particles pairs, the symmetry also predicts an electrically neutral W^0, which leaves a particle unchanged.

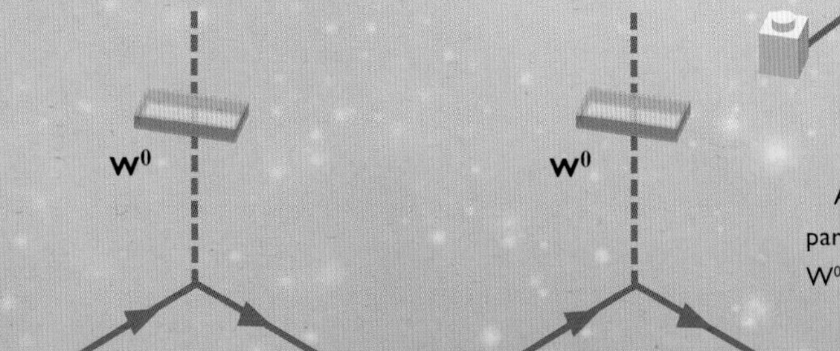

These beta-decay weak pairs are copied, just like the first generation of particles themselves, in the heavier quarks and leptons. After the discovery of the strange quark and development of quark theory, the weak force pairing led to the prediction and eventual discovery of the charm quark, the strange quark's weak charge partner. The top quark was also predicted and confirmed as the weak force partner for the bottom quark.

WEAK NUCLEAR FORCE

The weak nuclear force transforms particles from up-like to down-like quarks or from electrically charged leptons (e.g. electrons) to neutrinos. It is responsible for radioactive beta decay.

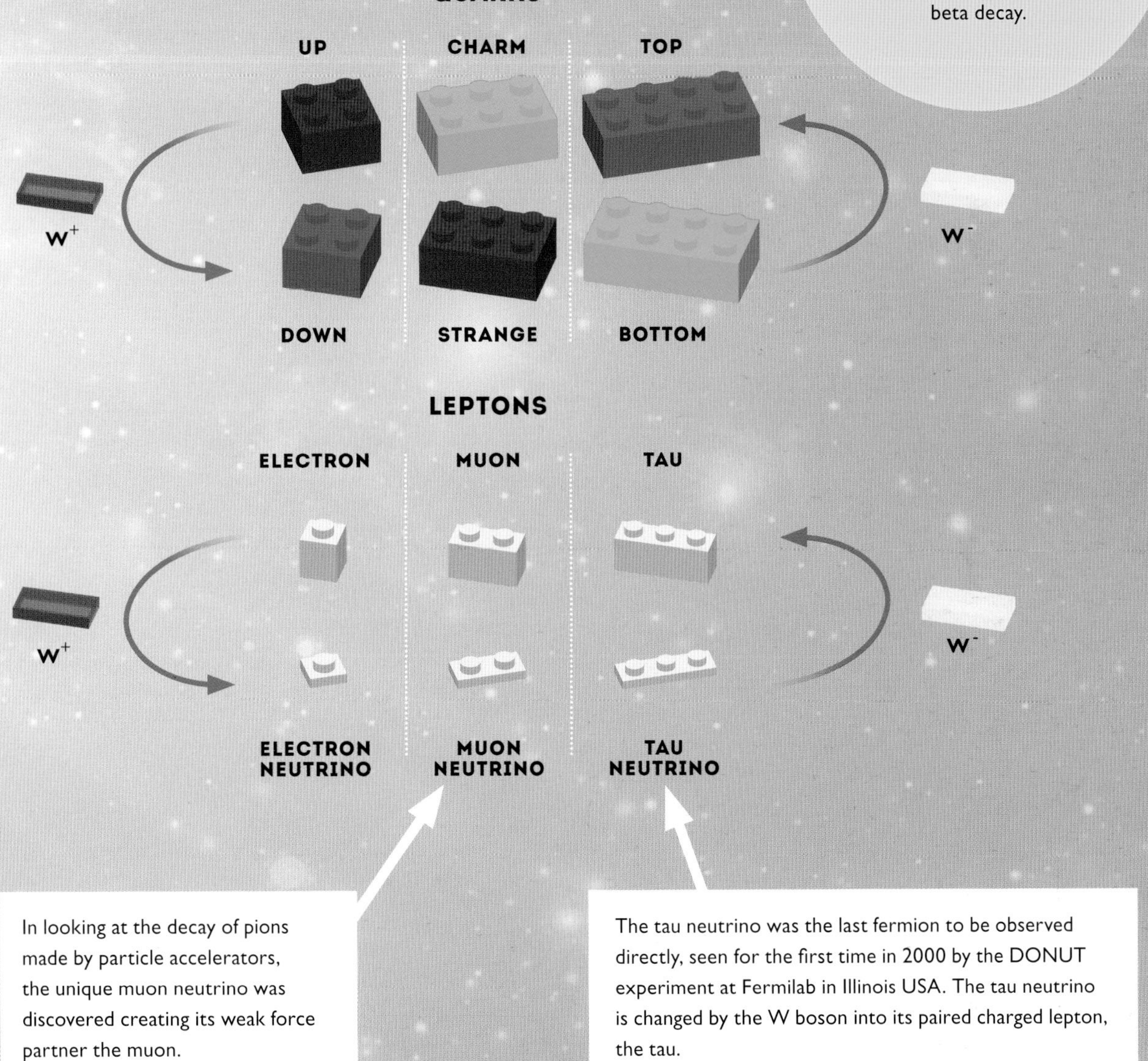

In looking at the decay of pions made by particle accelerators, the unique muon neutrino was discovered creating its weak force partner the muon.

The tau neutrino was the last fermion to be observed directly, seen for the first time in 2000 by the DONUT experiment at Fermilab in Illinois USA. The tau neutrino is changed by the W boson into its paired charged lepton, the tau.

Heavy particles wish to be as light as possible, and so they decay into lighter things

THE HEAVIER THEY ARE THE HARDER THEY FALL

The weak force is the reason that atoms as we know them exist. Without the W bosons swapping particles around in their pairs, the Universe would still be full of heavy particles like the top quark and the tau. Instead the weak force helps these particles lose mass – the heavier they are the more willing they are to flip to a lighter particle through the weak force.

Heavy tau particles do not hang around long, on average 0.3 millionths of a millionth of a second, before decaying to the very light tau neutrino. The W boson emitted when this happens then creates a new weak charge pair. As the tau is so much more massive than the neutrino, the W boson takes away a large amount of energy, which offers many options to what particle–antiparticle weak charge pairs it might create. It has energy enough to create a pair of muon and muon antineutrinos which it does roughly 17.4% of the time. It is also still able to create a pair of electron and electron neutrinos 17.8% of the time.

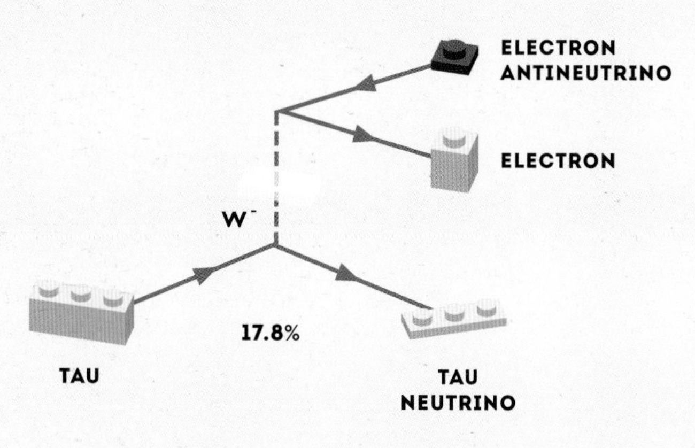

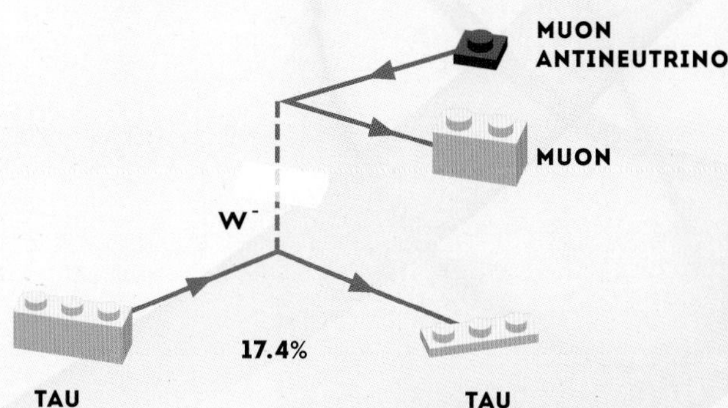

The tau is the only lepton massive enough to create quark–antiquark pairs, and therefore hadrons. It has energy enough to create multiple up and down quark–antiquark pairs. This creates single or multiple pions.

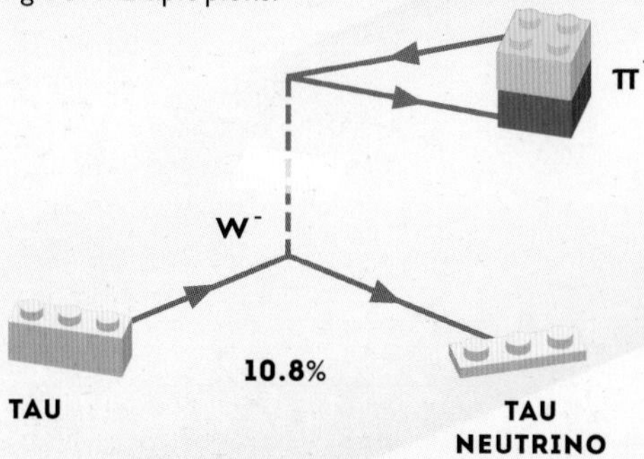

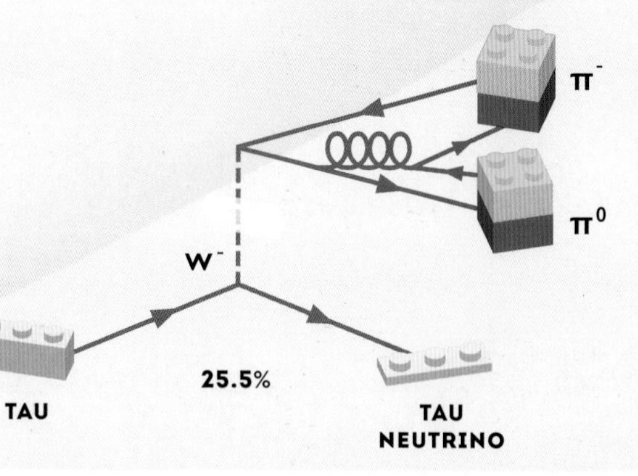

It is because of the wide variety of options that the tau is likely to decay quickly. More options to decay means it is more likely, just as the more building block sets are available, the more money I spend! Tau particles have average lifetimes of three tenths of a millionth millionth of a second, too quick to truly see them in a particle detector. They are only observed from the particles they decay into.

Muons decay to the muon neutrino with the emitted W boson taking away less energy. Less energy from a smaller difference between muon and muon neutrino mass means less options to create new particle–antiparticle pairs. In fact, the only real choice available to this W boson is to create an electron and its partner antineutrino.

Remember that you could switch all of the particles in the diagrams on this page for their antiparticle get the decay of the antilepton, for example, the antimuon.

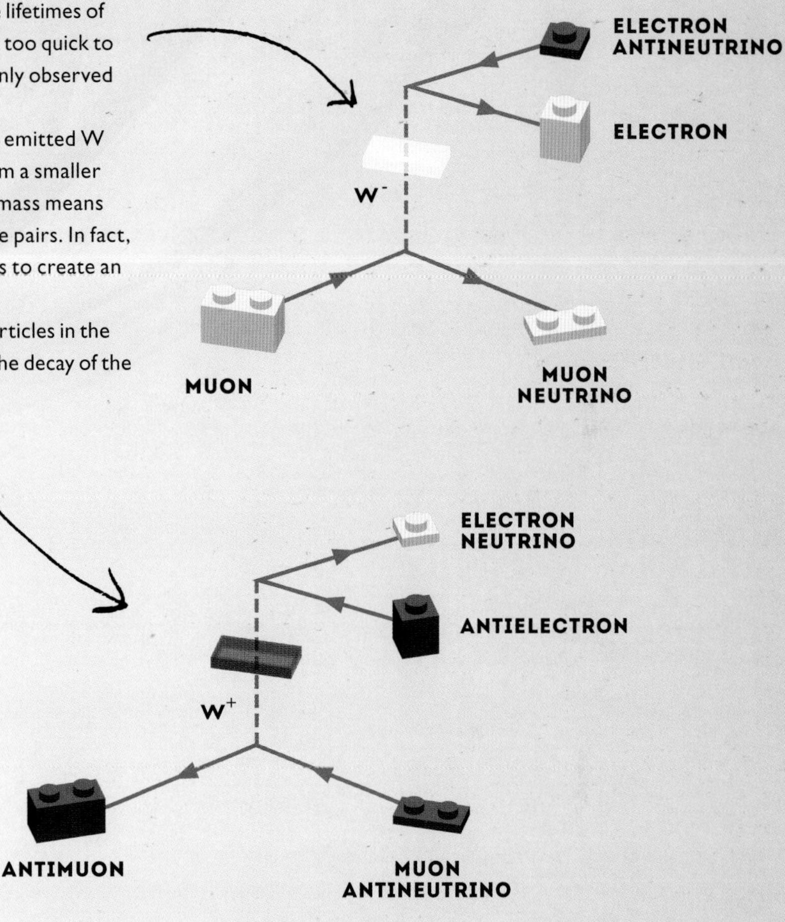

THE END OF THE LINE

Once at the electron there are no more options – an electron won't emit a W boson to change into an electron neutrino because the boson would have such little energy that it could not create any new particle–antiparticle weak pairs. The electron is stable and the end of the line, which is why it is found in all atoms and not the tau.

Intermingling of weak charge pairings of quarks is essential to forming protons and neutrons

QUARKS AND MIXING

After what I have said about leptons, the decay of quarks should make you worry – if the weak force only flips particles between the pairs, then there is a problem. All heavy top quarks, created just after the Big Bang, would decay to bottom quarks. These bottom quarks would then, like the neutrinos, be stuck as they cannot use the weak force to do anything but flip back to heavier top quarks. There should therefore be large numbers of particles made from the lightest quarks in the pairs: bottom quarks from top-quark decays and strange quarks from charm-quark decays. Instead, protons, neutrons and therefore over 99% of the Universe around us is made from just up and down quarks.

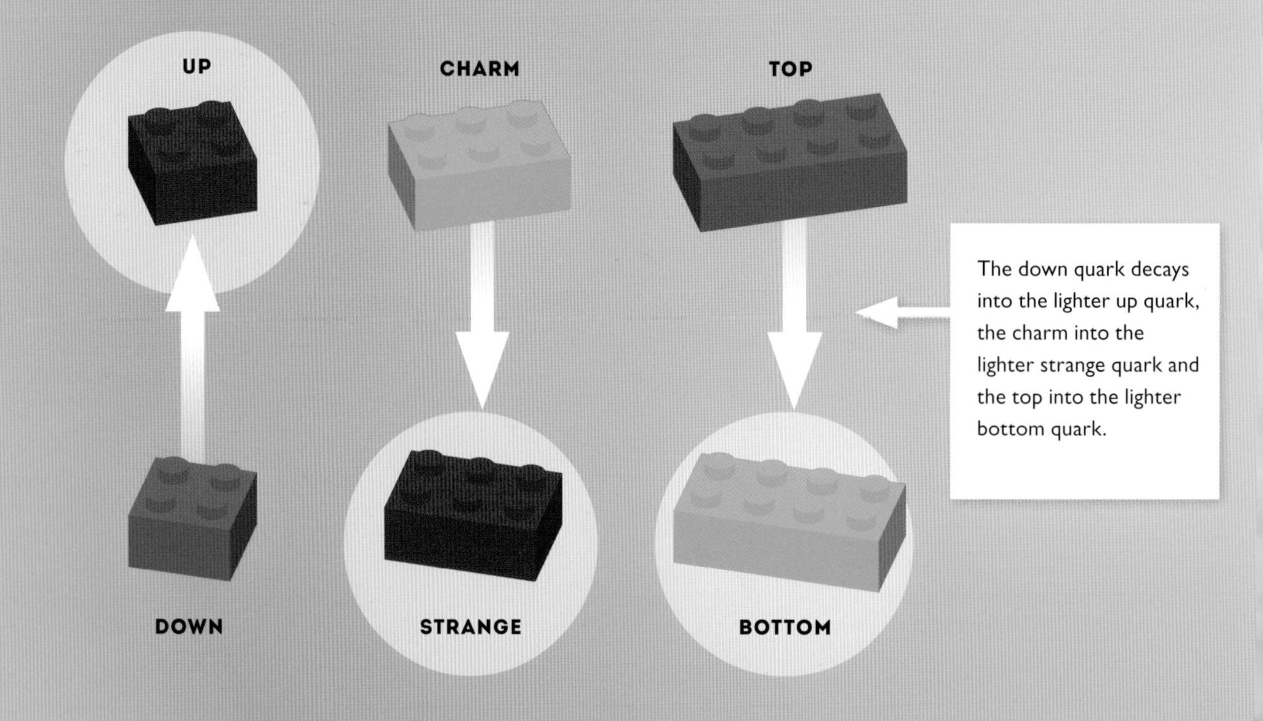

The down quark decays into the lighter up quark, the charm into the lighter strange quark and the top into the lighter bottom quark.

This is because quarks are not isolated in their pairs, but mingle and mix with each other. They do not do this when interacting with any of the forces, acting like brick particles, but instead when they are exploring all of their quantum paths (see pages 68–9 for a reminder). Each quark doesn't have just one but three pedometers, each representing a different way of measuring the path it takes. At the end of a quark's path the pedometers have all moved out of sync. The same way that comparing readings of pedometers from different paths determines the probability of where a particle will travel, comparison of the readings of the quark's three pedometers defines a probability that a quark can mix with a different weak-force pair (see page 147 for more).

If it can mix with another weak-force quark pair, then essentially the top quark has the possibility to decay not only to the bottom quark but also to the strange or down quark, although the last has yet to be confirmed. The probability of it doing this is determined by the amount that quarks are allowed to mix – something which can only be measured experimentally.

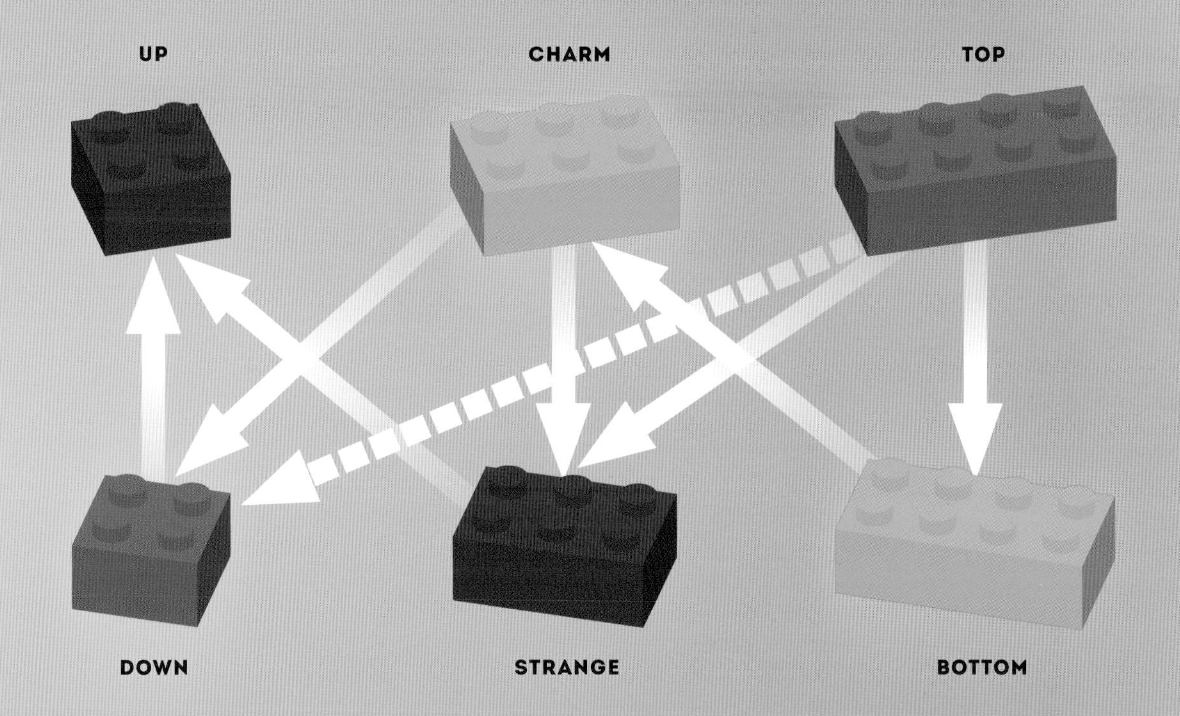

WHY?

Why quarks mix is, like much of the Standard Model, not understood any more deeply than 'because they can'. While it seems that leptons are fixed in their weak pairs it was discovered, through measuring ghostly neutrinos that this is not entirely the case.

QUANTUM

Quantum things are not solid ball-like particles but a collection of possibilities with different probabilities. It is this quantum ability of quarks that allows them knowledge of other possible ways in which they can decay, mixing up quarks between weak force pairs.

The weak force is only seen as weak because the particles exchanging the force cannot exist for long

WEAKER THAN USUAL

In Chapter 2, I said that the weak force started taking hold much much earlier than the electromagnetic force. Surely to overpower the energetic particles of the early Universe the weak force had to be stronger than the electromagnetic – well it is!

The coupling constant (which defines the probability that a particle will emit a boson) of the weak force is over five times larger than the 1/137 of the electromagnetic force. This means that particles should want to interact with weak bosons much more than they do with photons of the electromagnetic force. Yet a muon interacts with charged particles by emitting photons for some time before decaying by emitting a W boson.

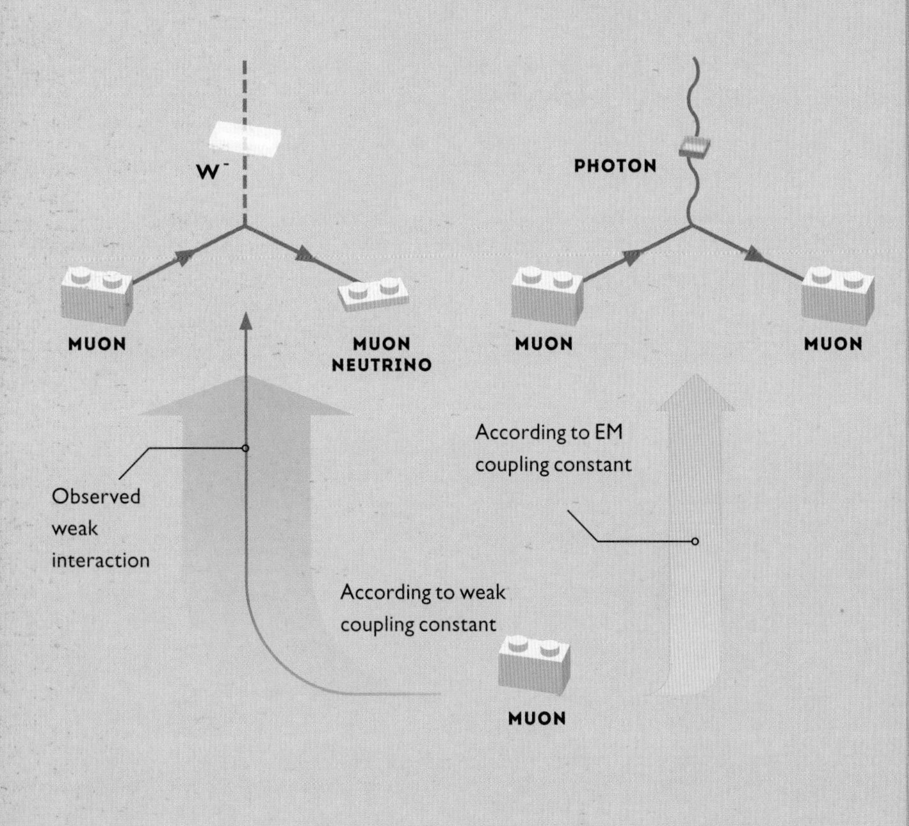

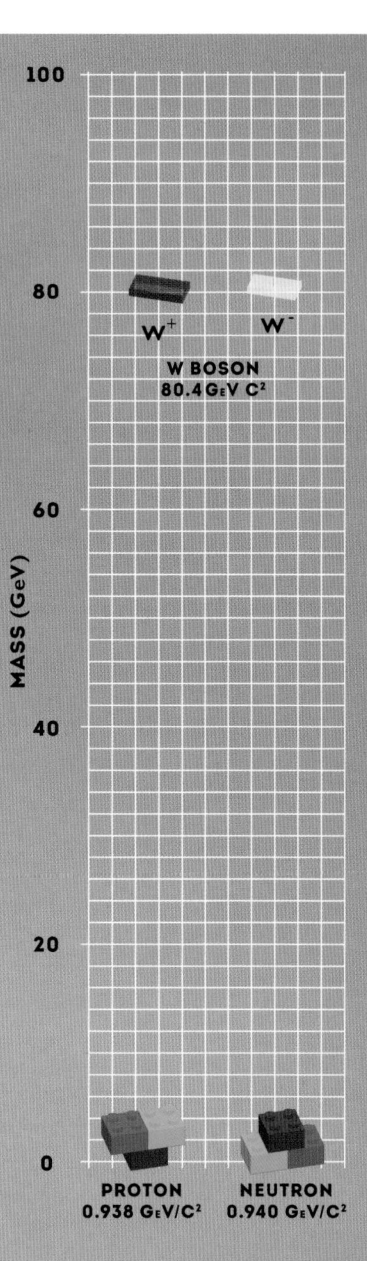

The reason the weak force appears so weak to us is because the W boson is massive – a whopping 85 times the mass of a proton! While a massless photon or gluon can be created with (almost) any energy, it requires a vast amount of energy just to create the mass of a W boson – after all, $E=mc^2$.

The tiny mass difference between neutron and proton could never create a W boson, so beta decay should never happen. The same holds for the mass difference between the muon and electron – the muon should not be able to decay. The reason weak decay does occur, however, is due to the same strange quantum behaviour which gives a proton most of its mass. The uncertainty principle mentioned earlier (page 116) explained that as we confine particles to ever smaller spaces their momentum increases. Looking at the Universe at smaller time frames, we find the accuracy of our energy measurements gets worse.

The delta, Δ, here shows uncertainty. As the uncertainty in the time *decreases*, as the line becomes more vertical, the uncertainty in the energy *increases*.

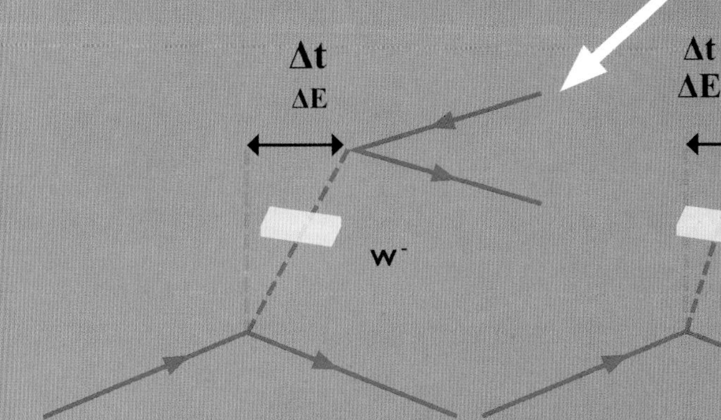

VIRTUAL PARTICLES

The more vertical the line in a Feynman diagram the shorter the time a particle exists for. As a particle's lifetime becomes less and less, the uncertainty in the energy the particle may possess increases thanks to the uncertainty principle.

Shrouded by this uncertainty, it is not against the laws of Nature to 'borrow' energy for short periods of time as long as it is quickly paid back. The tighter we squeeze time the more energy can be borrowed. A W boson can be borrowed only if it exists for about a billionth billionth billionth of a second, far too short a time to observe it. Because such particles can never be seen they have earned the name virtual particles and are drawn vertically in Feynman diagrams, existing in space but not in time. All other particles, which exist in space and time, are real particles which are possible to observe.

This short lifetime virtual behaviour decreases the likelihood that a particle has a chance of interacting with a W boson – the reason the weak force appears so weak.

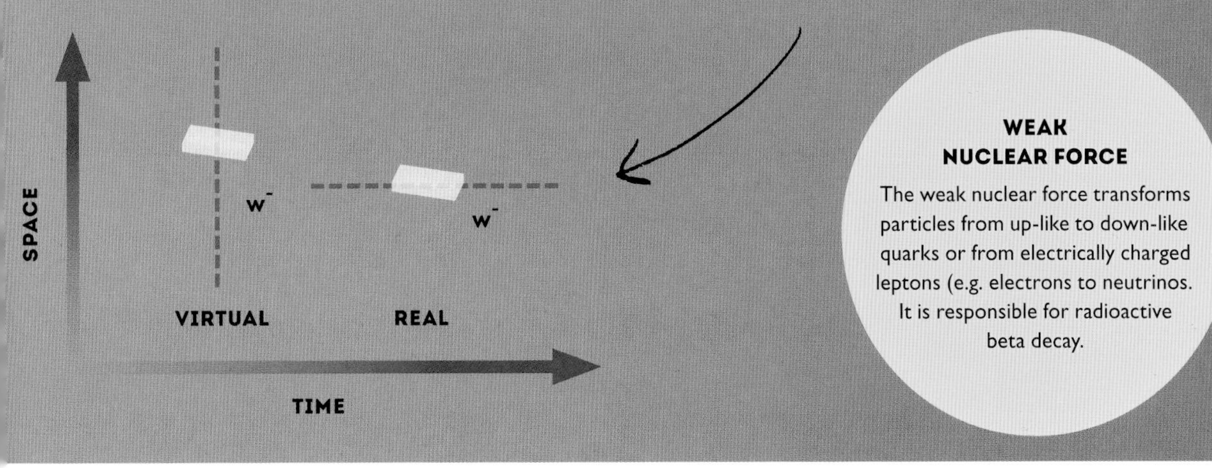

WEAK NUCLEAR FORCE

The weak nuclear force transforms particles from up-like to down-like quarks or from electrically charged leptons (e.g. electrons to neutrinos. It is responsible for radioactive beta decay.

In the early Universe the electromagnetic and weak forces merged into the electroweak force

NOT SO WEAK AT EARLIER TIMES

This weak force wasn't always so weak. One hundred billionths of a second after the Big Bang, particles in the hot dense Universe had an average energy around 100 GeV, enough to create a W boson without the need to borrow additional energy. These real and not virtual W bosons existed for no more than a fleeting moment. At this time the electromagnetic and weak forces would have appeared essentially identical to one another with particle–antiparticle weak pairs forming W bosons the same as particle–antiparticle pairs form photons in annihilation.

ELECTROWEAK FORCE

Just like identical twins, if there is no way of telling the difference between two things from the way they interact, then there must be a symmetry at work. The symmetries representing the electromagnetic and weak forces are combined at these high energies to create a new symmetry representing a single unified electroweak force.

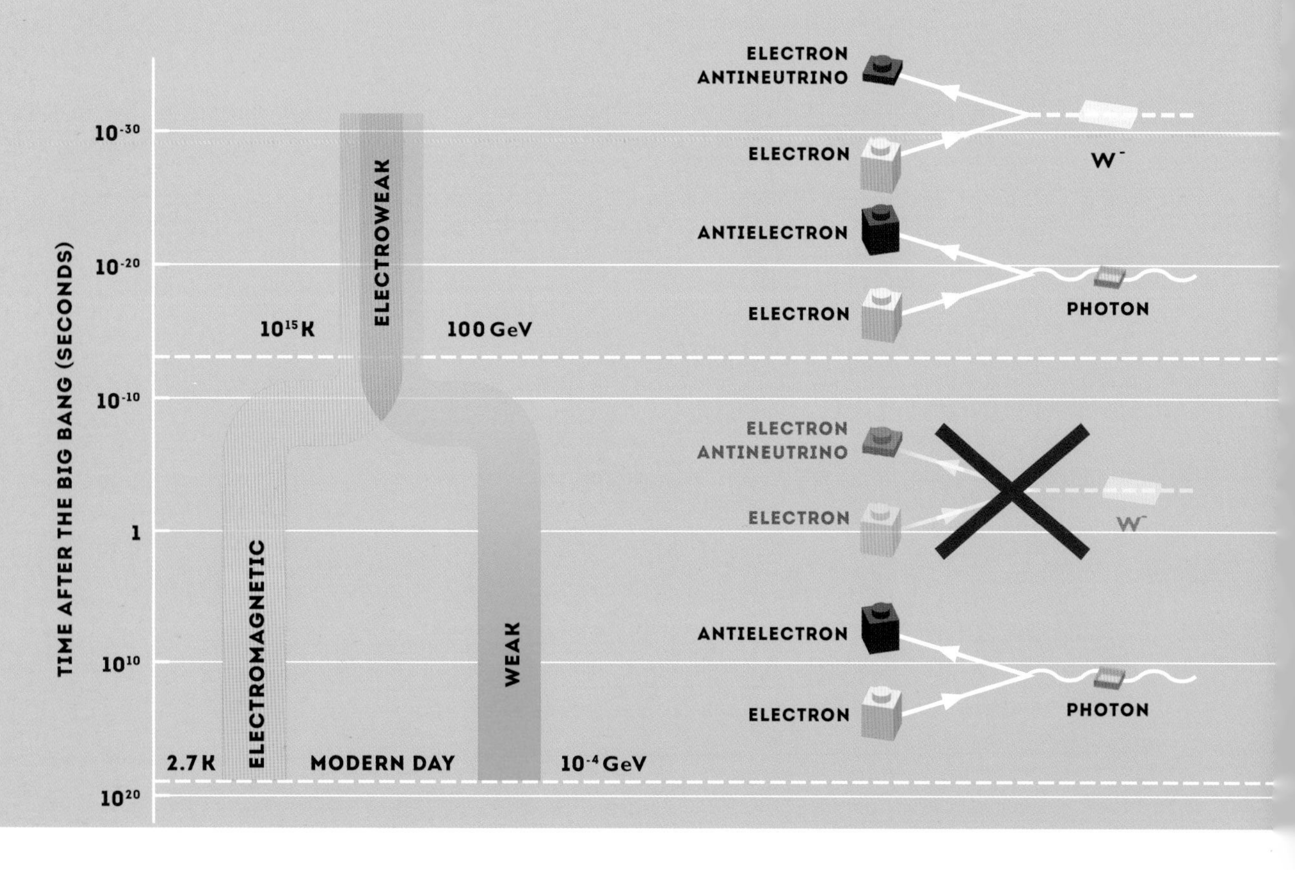

When combined into the electroweak force, the W^+ and W^- crystallize out as distinct particles because they have unique interactions, flipping particles one way or another. The W^0, however, behaves very similar to a photon – you can switch one for the other in some Feynman diagrams and get the same outcome.

As with all other indistinguishable particles, like the different quark configurations of the π^0 (see page 104), the behaviour of the particles becomes mixed together. The true boson of the pedometer symmetry I mention in chapter 3 is in fact not the photon, but a B boson. It is this B boson which is indistinguishable from the W^0. When the two mix, they form two new particles. Each a combination of a bit of B and a bit of W^0. The massless photon, as the measurable boson of the electromagnetic force, is one. The other is called the Z^0 which like the W bosons, but unlike the photon, has a mass.

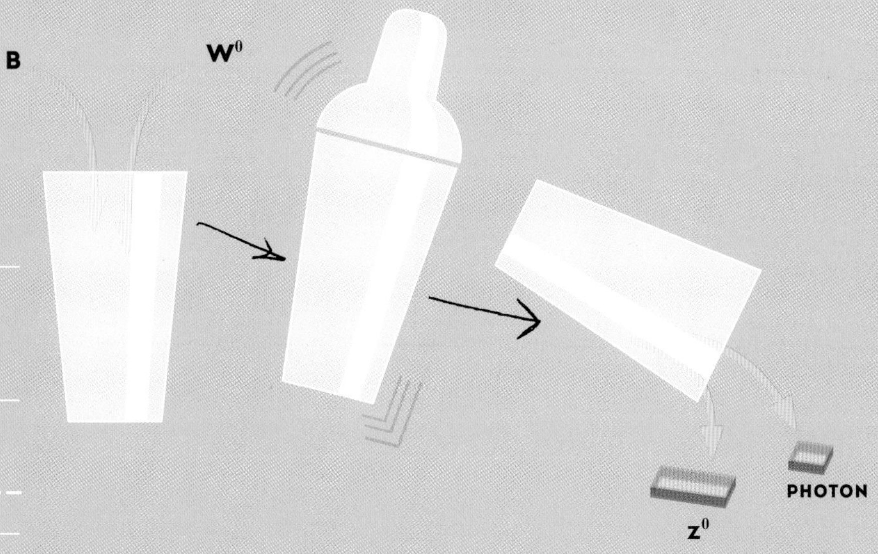

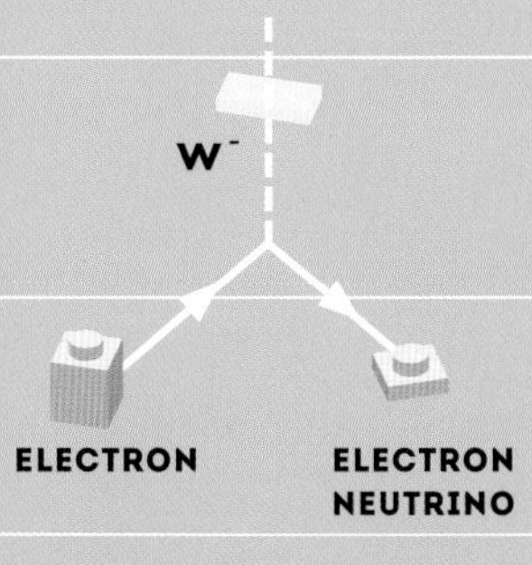

B
W^0
PHOTON
Z^0

W^-
ELECTRON
ELECTRON NEUTRINO

As the Universe expanded and cooled, particle energies soon dipped low enough that they were no longer able to create W or Z^0 bosons without borrowing energy. Weak and electromagnetic forces started to take on very different characteristics, as W and Z^0 bosons could only be exchanged virtually. The symmetry describing the electroweak force was broken by mass into two forces. Breaking this symmetry and giving mass to the weak bosons required a new type of force unlike the others we have encountered so far.

There are only three generations of particle that interact through the weak force

JUST THREE GENERATIONS

The Z^0 is the only boson which can decay into a pair of neutrino and antineutrino which leave most particle detectors unseen. It is important to understand the likelihood of a Z^0 decaying in this way.

Like any massive particle, the W and Z weak bosons decay, forming new lighter particles. As we have seen from beta decay above, the W bosons produce weak-charge pairs of particle and antiparticle. With a mass of around 80 GeV/c^2, a W boson cannot create a pair containing a top quark as it is far too massive. Other than this, quark mixing means a W can decay into all other combinations of pairs of up-like and down-like, quark and antiquark.

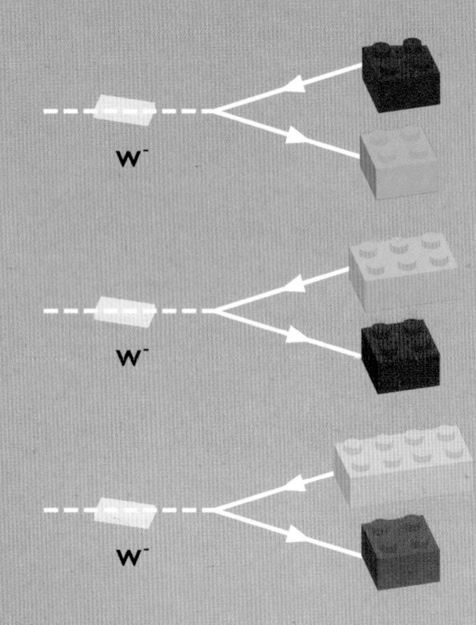

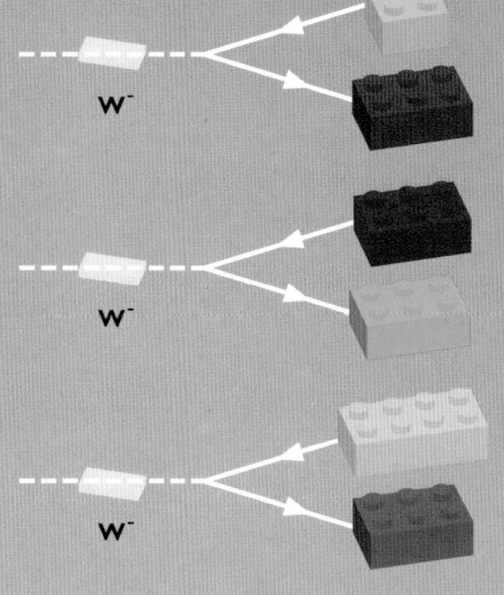

The W boson can also decay into particle–antiparticle pairs of charged lepton and neutrino.

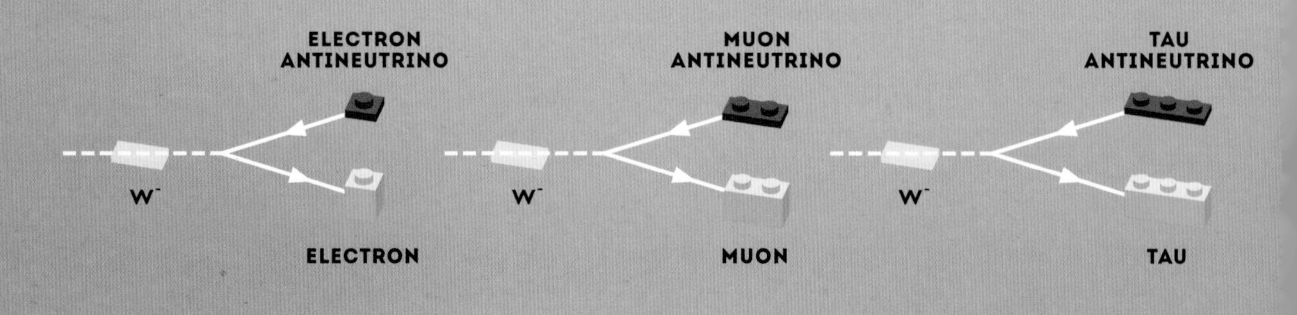

The Z^0 boson, really a heavy version of a photon, creates pairs of particle and antiparticle. It has mass enough to create pairs of all but the top quark and antiquark.

Unlike the photon, however, the Z^0 can also couple to neutrinos. A Z^0 can produce pairs of any charged lepton and its antiparticle or any neutrino and antineutrino.

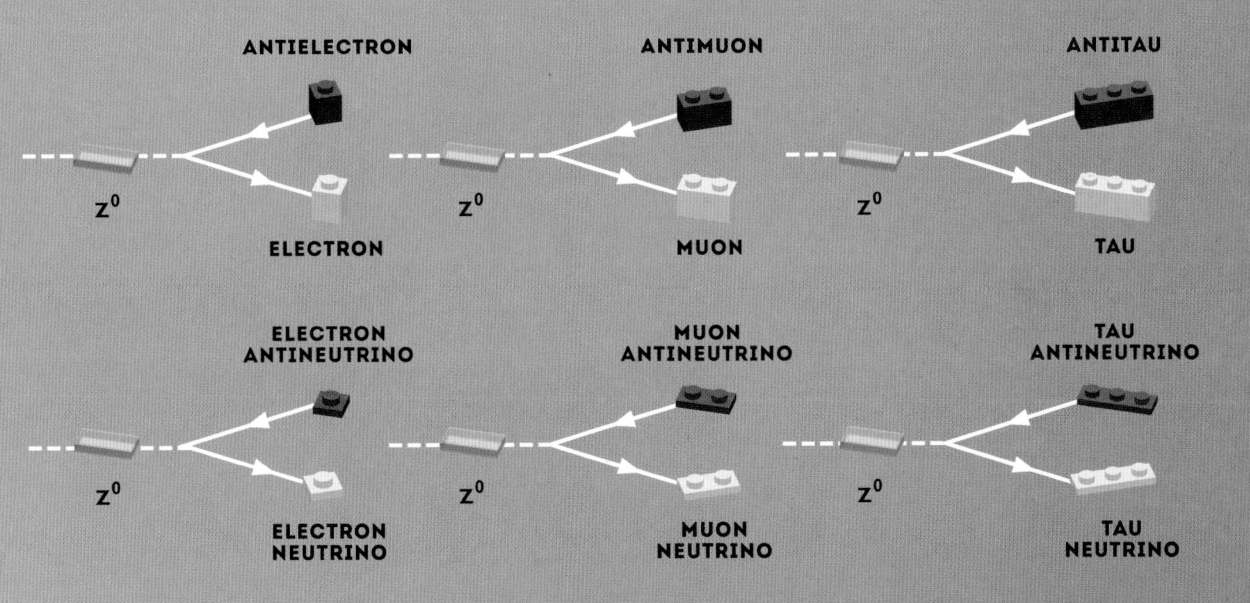

At the Large Electron Positron (LEP) collider, the predecessor to the LHC in the 27km tunnel at CERN, W and Z bosons were produced on an industrial scale. With so much data the ALEPH detector (Apparatus for LEP PHysics) was able to measure the average lifetimes of the Z^0 which decayed in each of these different ways. When comparing the decay of Z^0 into charged leptons and neutrinos, they determined that there could only be three generations of neutrino. If there were more generations, then there would be more options for the Z^0 to decay to, which would dramatically reduce its lifetime. So we can say with marked certainty that there are no more than three generations of neutrino unless, that is, the fourth generation of neutrino is billions of times more massive than the others – greater than half of the Z^0 mass at over 45 GeV.

WEAK FORCE AND BREAKING SYMMETRIES SUMMARY

The weak force is combined with the electromagnetic to form the single electroweak force at high energies. We see weak and electromagnetic as two different forces today because the weak-force bosons have a mass and so behave very differently at such low energies.

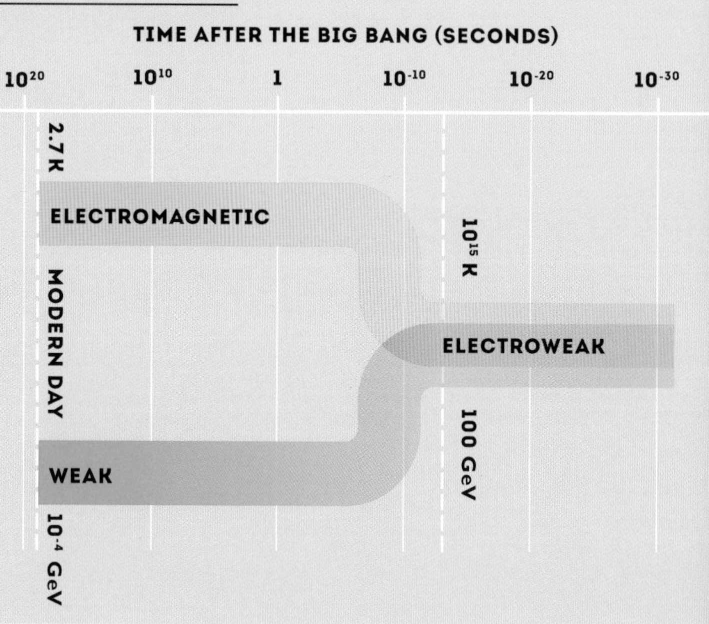

Weak charge is two-dimensional, grouping particles into pairs like different sides of a coin.

There are three weak-force-carrying boson particles; the W^- and its antiparticle W^+ version, and the Z^0 (which is its own antiparticle).

The W bosons flip the weak charge, changing a particle into its weak-paired partner; up-like quarks to down-like quarks and charged leptons to neutrinos – to do this they have to take away electric charge.

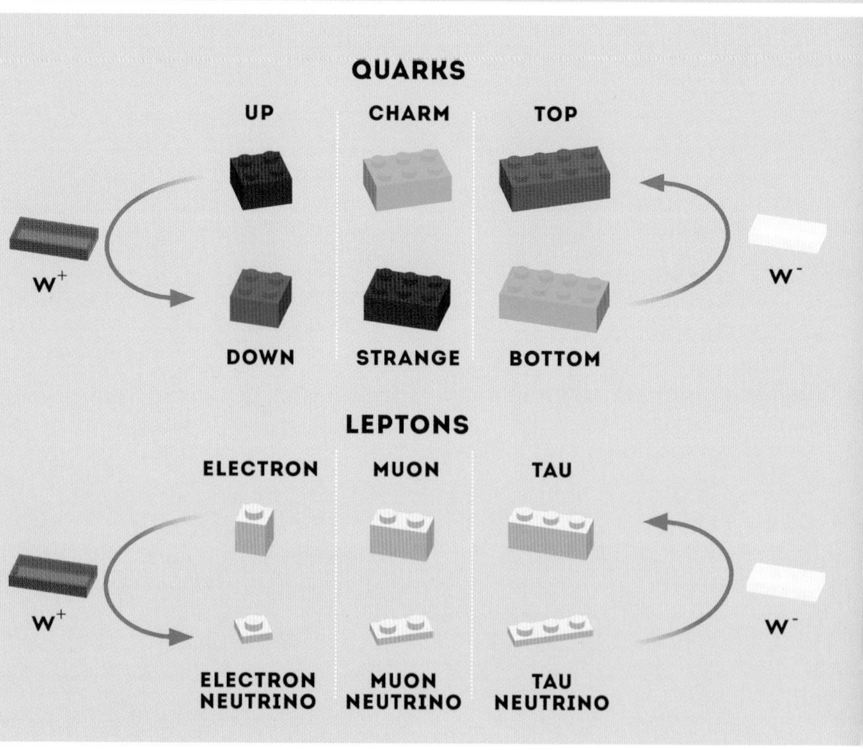

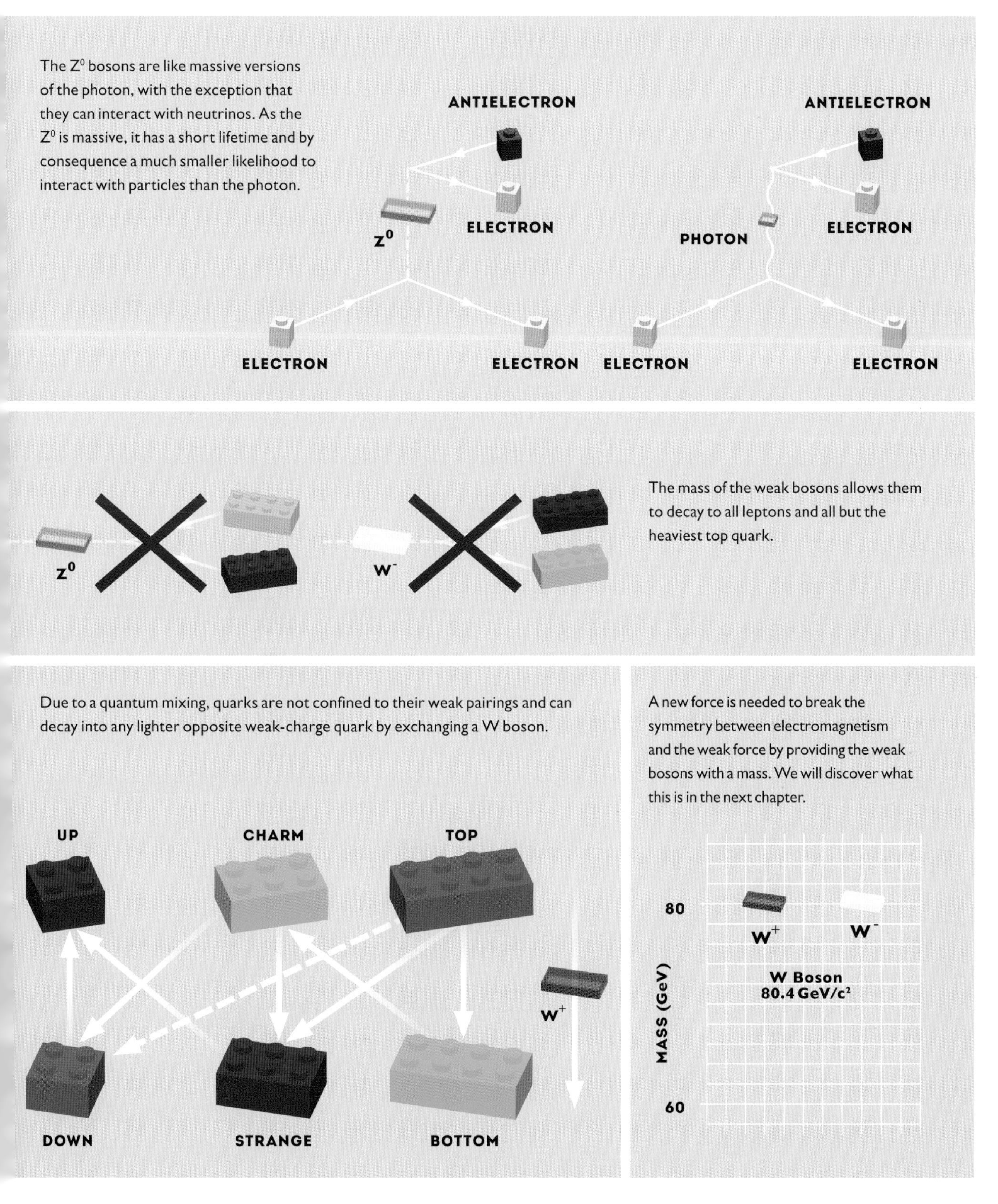
The Z^0 bosons are like massive versions of the photon, with the exception that they can interact with neutrinos. As the Z^0 is massive, it has a short lifetime and by consequence a much smaller likelihood to interact with particles than the photon.
ANTIELECTRON
ELECTRON
Z^0
ELECTRON
ELECTRON
ANTIELECTRON
PHOTON
ELECTRON
ELECTRON
ELECTRON
Z^0
W^-
The mass of the weak bosons allows them to decay to all leptons and all but the heaviest top quark.
Due to a quantum mixing, quarks are not confined to their weak pairings and can decay into any lighter opposite weak-charge quark by exchanging a W boson.
UP
CHARM
TOP
W^+
DOWN
STRANGE
BOTTOM
A new force is needed to break the symmetry between electromagnetism and the weak force by providing the weak bosons with a mass. We will discover what this is in the next chapter.
80
W^+
W^-
W Boson
80.4 GeV/c²
MASS (GeV)
60

Particles get their mass though interactions with the Higgs force field, via the Higgs boson

A NEW FORCE

Force fields of the electromagnetic, weak and strong forces have a limited range of influence. The field of the fourth force which gives mass to the weak bosons must, however, extend across the whole of space. If it did not, then the bosons would have different masses in different parts of the Universe. The force was suggested by a number of physicists almost simultaneously, but has become known as the BEH field – after Robert Brout, François Englert and Peter Higgs. Higgs and Englert were awarded the 2013 Nobel Prize in Physics for their work. Brout missed out having passed away in 2011.

Reluctant to step into the spotlight since he was announced as a recipient of the 2013 Nobel Prize in Physics, Professor Peter Higgs has proved to be as elusive as the particle which bears his name.

Any particle with a mass gains that mass by interacting with the BEH force field. Interactions with the field are exchanged, like all of the other forces, by a boson particle. The boson of the BEH field was first suggested by Peter Higgs, as an after-addition to his paper to get it published, and so it is commonly just called the Higgs boson. Not only does it interact with all fermions and the weak bosons but it also interacts with itself – how else would it have a mass of 125 GeV/c^2?

Higgs bosons interact with particles, diverting their otherwise light-speed journey, seeming to slow it down. Einstein's Special Relativity tells us that the reason an object cannot reach light speed is because it has a mass.

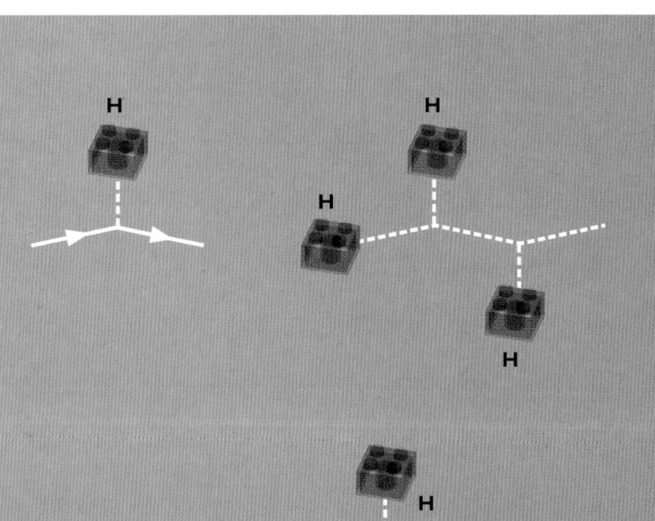

The more a Higgs boson interacts with a particle, the greater the slowing of its path and therefore the greater a particle's mass. It is through interactions with the Higgs field that a W boson gains a mass and at low energies breaks the symmetry of the electroweak force into the two electromagnetic and weak forces we experience today (pages 132–3).

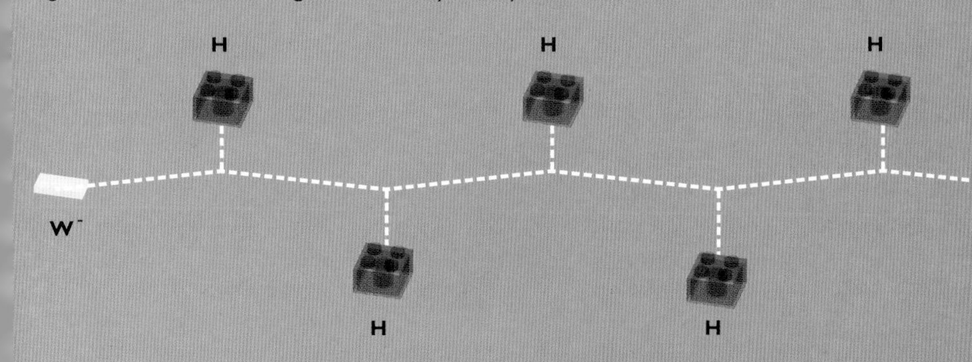

A breaking of symmetry like this can be thought of as marking one stud on a plastic brick. Before the brick would have been symmetric under rotations of 90°, but this marked brick does not return to the same state it was in – it is clearly different. The BEH field does something similar to this, marking bricks with a mass and breaking, what is at high energies, the electroweak symmetry.

BOSONS

Bosons are the particles which exchange the forces of Nature, carrying energy and charge between particles.

There are many ways to create a Higgs boson, and even more ways for it to decay

HUNTING A HIGGS

So if the BEH field is everywhere, why do we not see Higgs bosons all the time? It is because they, like the W and Z bosons, are massive and so it takes a lot of energy to create them. At low everyday energies they only exist as virtual particles. To detect a Higgs boson directly, we have to strike the right energy to bring a virtual Higgs boson into the real world. This is done at the Large Hadron Collider in four main ways, each involving the annihilation of a particle and an antiparticle.

1

Gluon fusion: two gluons create top or bottom quarks and antiquarks which eventually annihilate one another to form a Higgs boson.

2

Top quark fusion: a top quark and an antiquark (created by gluons) annihilate one another to form a Higgs boson.

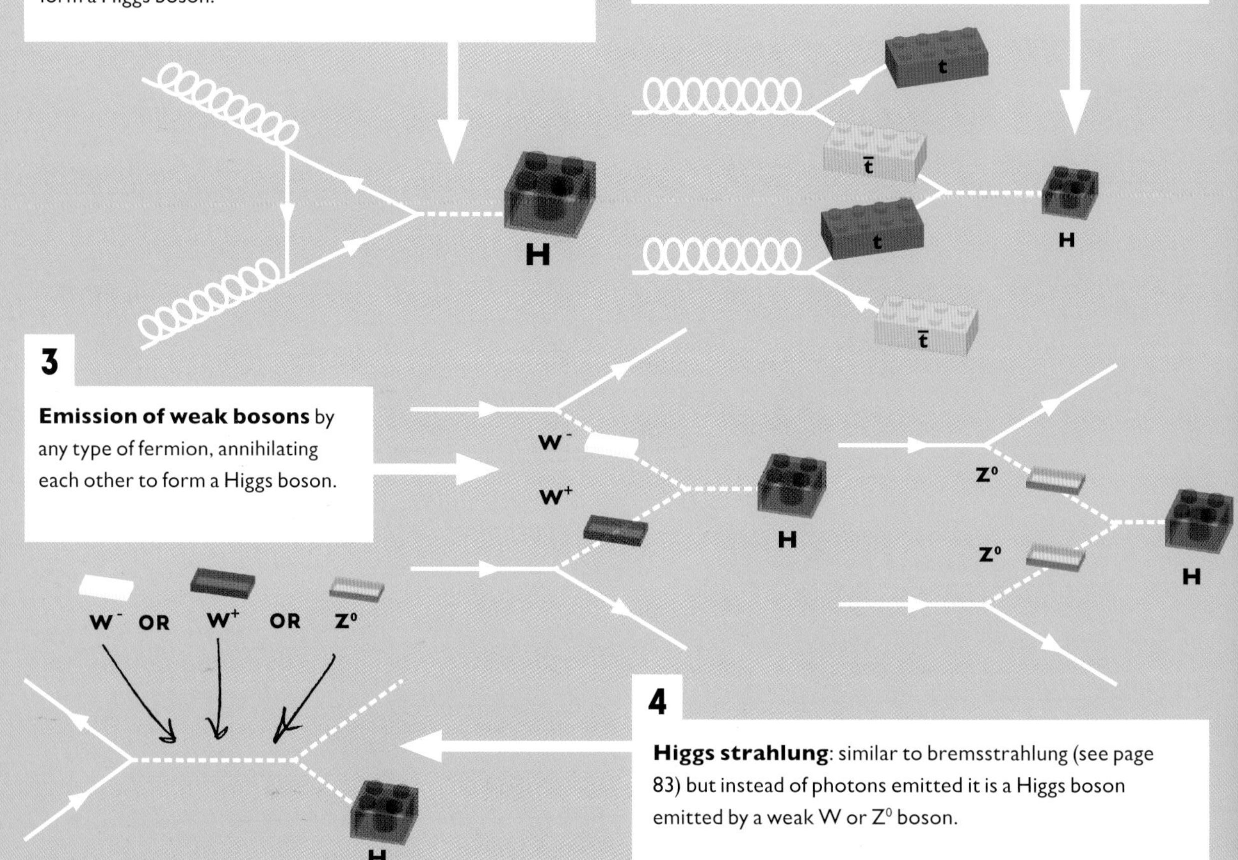

3

Emission of weak bosons by any type of fermion, annihilating each other to form a Higgs boson.

4

Higgs strahlung: similar to bremsstrahlung (see page 83) but instead of photons emitted it is a Higgs boson emitted by a weak W or Z^0 boson.

Once created, the short-lived Higgs quickly decays into a particle–antiparticle pair. The existence of the Higgs can only be inferred, like the discovery of the W and Z bosons, from summing the energy of the decay particles seen. If there is an excess of measurements at a particular summed energy which cannot be explained from the other particles we are aware of, and it is significant enough, then it can only be from the decay of a new particle.

Higgs bosons interact with any particle that has a mass and therefore decays into any bosons or fermions with a mass – the more massive the particles the more likely a Higgs will decay into them.

The following diagrams are the most likely ways that a Higgs boson might be observed decaying in either the ATLAS or CMS detectors at the Large Hadron Collider.

1

The reverse of the gluon fusion results in the top or bottom quark and antiquark annihilations forming photons.

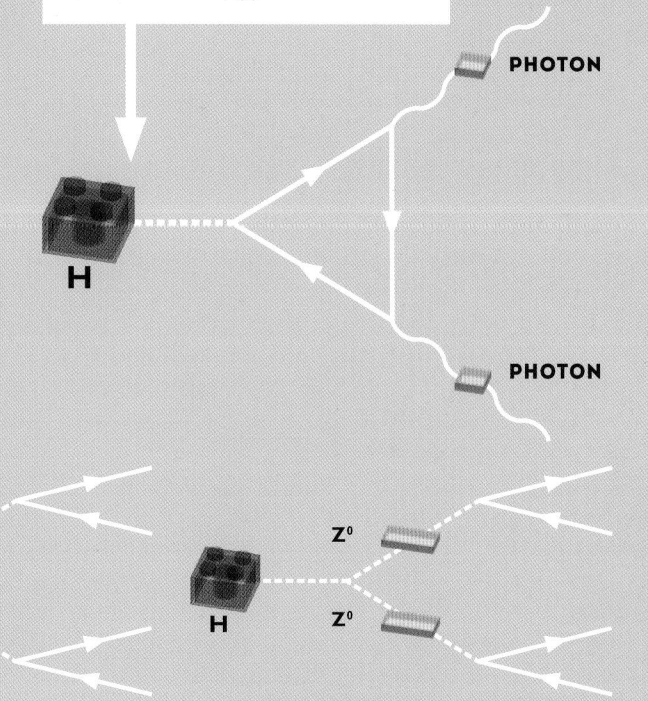

2

The reverse of the weak boson annihilations eventually produces two leptons and two antileptons.

3

The Higgs can also decay directly into fermion particle–antiparticle pairs. Most likely are the massive tau leptons, as seen below.

4

Also high on the list is the decay into a bottom quark and antiquark. Although this is yet to be confirmed, the others on this page have been verified at the LHC.

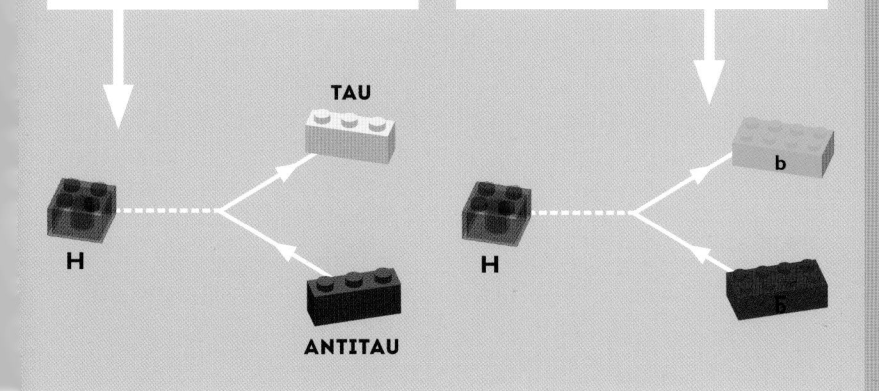

FINDING THE HIGGS

In 2012, almost half a century after it was first predicted, the ATLAS and CMS experiments at the LHC independently presented the evidence everyone had been waiting for – a bump in data which could only come from a new particle. The few years that followed saw confirmation that this was indeed a Higgs boson, but is it the only one? (See pages 166–7.)

Particles come in left and right and interactions with Higgs bosons swaps between the two

LEFT AND RIGHT

The electromagnetic, weak and strong forces all influence the energy, charge or type of particle during interactions, but on the face of it the Higgs leaves particles untouched. To explain more fundamentally what the Higgs does, we need to talk about another property of particles called chirality, from the Greek word for hand.

Particles can come, like hands, in two similar but distinguishable left-handed and right-handed versions. Your right and left hands are mirror images of each other. Both are 'hand shaped', but there is no rotation or movement that could make them look exactly identical. Up until this point we have distinguished our brick particles by properties like their shape and colour. This is enough for the strong and electromagnetic forces, but the symmetry-breaking of the weak force by the BEH field leads to this new defining characteristic.

The characteristic is so subtle that we have to look at the writing on the studs to notice it. Take, for example, the word BRICK printed on every plastic brick. If we truly reflected an image of a brick in a mirror, the writing on the stud would read ꓘƆIЯB. Only when we look at the studs closely can we determine if a brick is left-handed or a mirror right-handed version.

BRICK

LEFT-HANDED

ꓘƆIЯB

RIGHT-HANDED

The writing on the stud does not affect how you build neutral atoms through electromagnetic attraction or hadrons through the binding strong force – the studs still connect the bricks together. The strong and electromagnetic forces are ambidextrous in that both right- and left-handed particles interact with them – the same way we are happy to pass objects to another person with either of our own hands.

Electrically positive nuclei and negative electrons will attract one another no matter what their chirality (orientation of writing on the studs). They will still happily form electrically neutral atoms such as helium (top) or hydrogen (bottom).

The effect of the BEH symmetry breaking, however, breaks the ambidexterity of the weak force. Before symmetry breaking, the interaction of all of the forces can be thought of as a high-five between two hands, they can interact no matter if they are left or right handed. After symmetry breaking the strong and electromagnetic forces remain ambidextrous like a high five. The weak force, however, changes its interaction to be more like a hand shake, where a left hand can only shake another left hand and a right hand only another right hand. The weak force also only allows left-handed particles to interact. Right-handed particles cannot interact through the weak force at all.

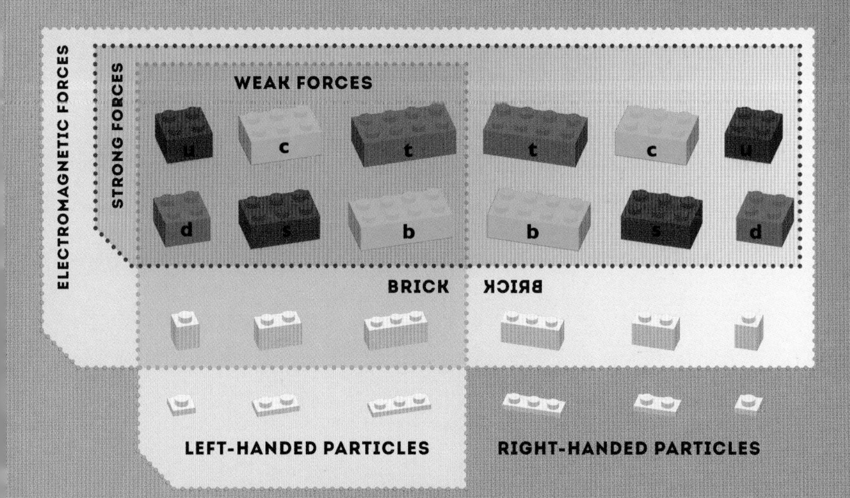

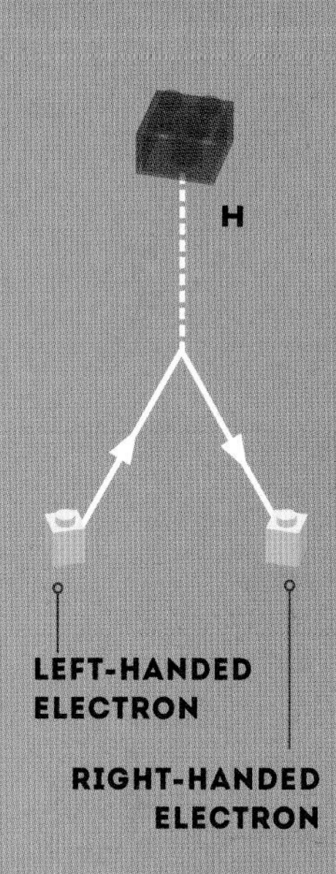

When the weak force interacts with antiparticles it does the opposite. It only interacts with right-handed antiparticles, bricks which have a mirrored ꓘƆIЯꓭ printed on the studs and not the left-handed BRICK.

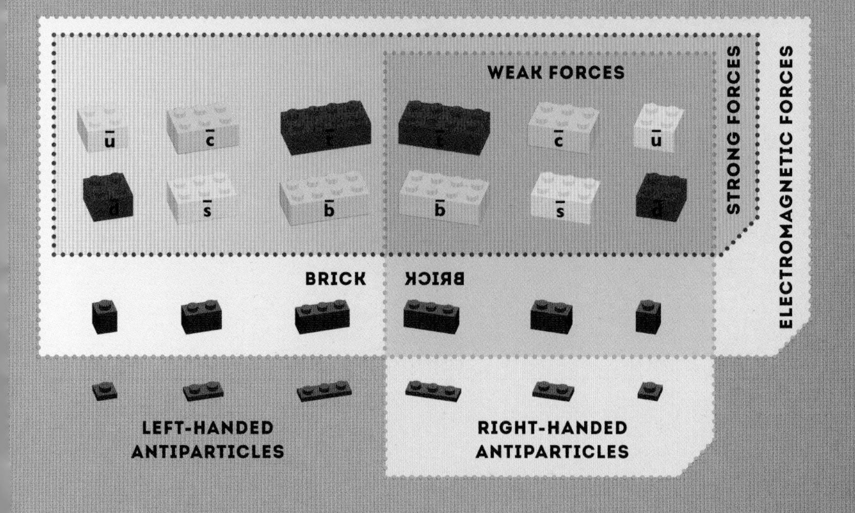

Interaction between a fermion and a Higgs boson flips the chirality of the particle, like mirroring the logo on the studs, from left-handed to right-handed. It is this new distinguishing feature which leads to the weak force breaking symmetries, like strangeness (see page 97), which are otherwise respected by the electromagnetic and strong forces.

Despite being the most plentiful particle, we have to use large detectors to see any neutrinos

GHOST HUNTING

Despite the unimaginable number of atoms in our Universe, each made from collections of up and down quarks and electrons, there is no fermion more numerable than neutrinos. They were produced in vast numbers just after the Big Bang (see page 40) as the weak force transformed protons to neutrons.

In every cubic centimetre of space, about the size of the tip of your thumb, there are twelve neutrinos from this early time – add up every neutrino across all of space and you get a billion times more neutrinos than all other fermions combined.

Neutrinos are still produced in large numbers in the same way inside every star today (pages 46–7). Hundreds of millions pass right through your body every second, released from the fusion process in the Sun. They are also released in the beta decay of radioactive elements or in the decay of any heavy particle where a W boson forms a charged lepton.

Most neutrino experiments detect these ghostly particles from the charged leptons they transform into when exchanging a weak W boson. The charged lepton is usually high in energy and travelling faster than light can within the water, resulting in the creation of Cherenkov radiation (see pages 78–9). Meanwhile the emitted W boson transforms a down quark in a neutron into an up quark turning it into a proton.

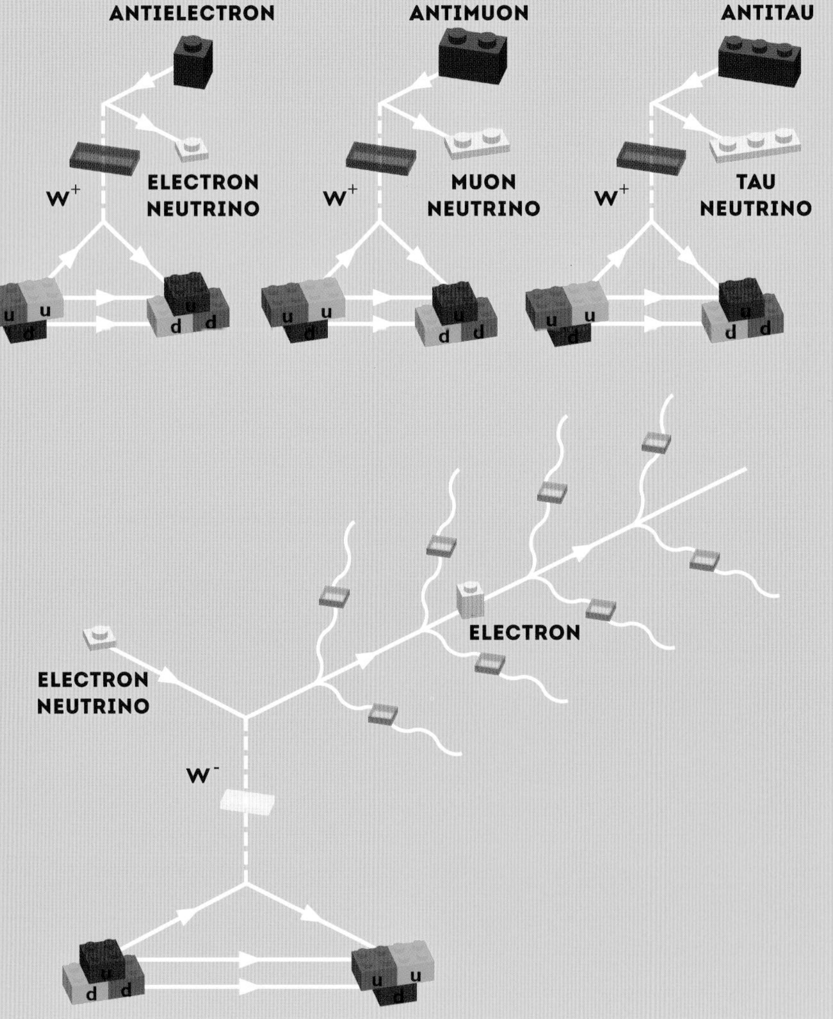

Despite being so plentiful, neutrinos remain mysterious thanks to their reluctance to interact with the world around them. With no electric charge, their only option to do anything lies in the improbable weak nuclear force. Detecting a neutrino is therefore a game of chance. To increase the likelihood of seeing a neutrino in a detector, you need more stuff for neutrinos to interact with, which has led to the world's largest particle detectors.

Most detect the charged leptons produced when the neutrino exchanges a W boson. Water Cherenkov detectors sense faint Cherenkov radiation from these fast-moving electrically charged particles (see pages 78–9). The Super Kamiokande (Super-K) detector in Japan is the largest of such detectors still running today, containing some 50,000 tonnes (over 33 full Olympic sized swimming pools) of ultra-pure water. The Sudbury Neutrino Observatory (SNO), which ran from 1999 to 2006, also used this technique.

SNO could also detect neutrino interactions involving a Z boson. It contained heavy water – in place of hydrogen was a hydrogen-2 isotope called deuterium. A Z boson exchanged by a neutrino would split apart the proton and neutron in the deuterium nucleus. A small, but detectable, amount of light would then be released when the neutron was absorbed by another nucleus.

The Sudbury Neutrino Observatory (SNO)

NEUTRINOS AND ANTINEUTRINOS

OR

OR

The Z^0 from interacting neutrinos breaks apart the deuterium nucleus into free a proton and neutron.

Z^0

Neutrinos suffer from a crisis of identity and change type over large distances

IDENTITY CRISIS

Originally the Standard Model assumed that neutrinos were massless, as there was no evidence that they had or required a mass, only that they take away a share of energy in weak interactions like beta decay. However, the more neutrinos that were observed, the stranger their behaviour appeared.

The number of electron neutrinos copiously produced as a product of fusion processes inside the Sun (see pages 46–7) can be predicted from knowledge of the weak force and the size and brightness of the star. As experiments recorded sightings of these solar neutrinos, it soon became clear that fewer than expected were being seen – around one third.

A similar discrepancy between prediction and observation was also seen in muon neutrinos produced in cosmic ray showers (see page 95). Despite expecting an even distribution of neutrinos coming into the detector, the number which entered from below, which had journeyed through the Earth, was half of that raining down from above. Half the muon neutrinos could not have just disappeared as a block of lead around 100,000 light years thick would be needed to stop that many neutrinos. So where were they going?

Neutrinos were changing into different types, but they did not have enough energy to create the heavier charged leptons associated with the neutrinos they had become. Still, both results showed that neutrinos seemed to be disappearing after longer journeys. As they are the lightest particle known, decay to other particles was not an option. Instead what was happening, determined by both the Super-K and SNO experiments, was very similar to the quark mixing described on pages 128–9.

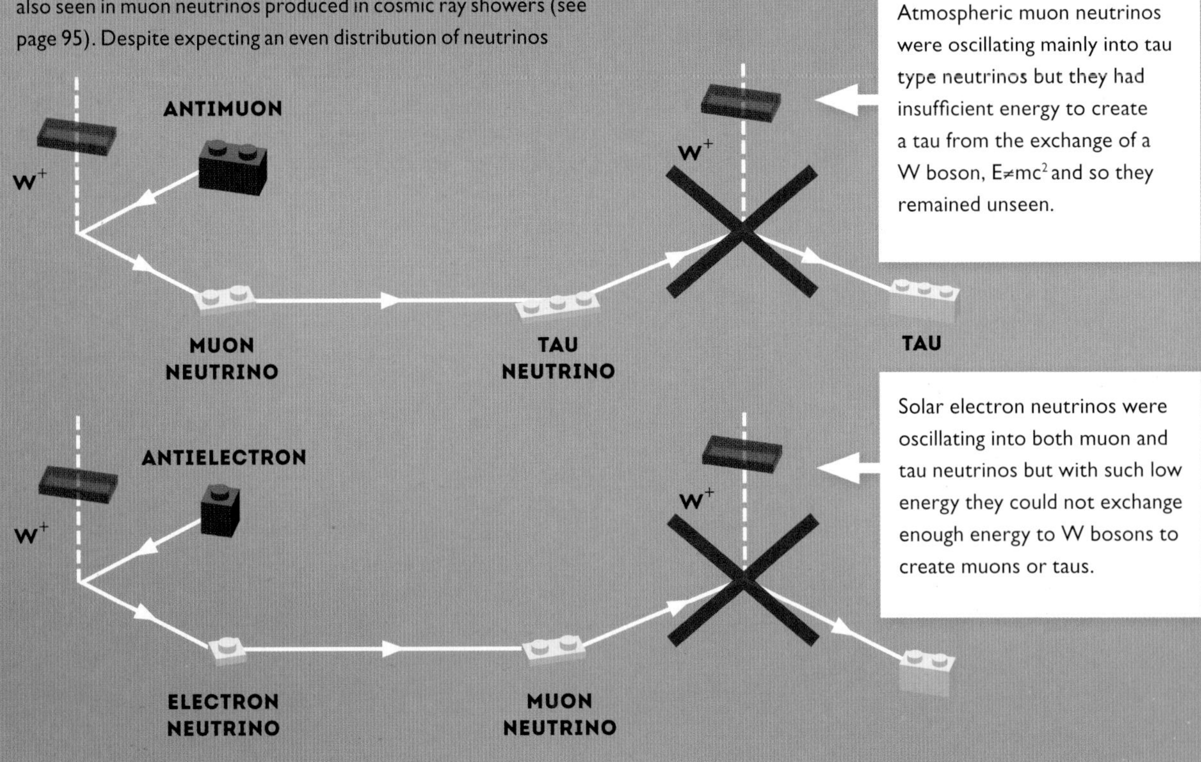

When produced, a neutrino's internal pedometers were set to identify it as 100% electron, 100% muon, or 100% tau and 0% of the others. As a neutrino travels, these pedometers turn at different rates. If compared at a later time they no longer represent the same percentage likelihood of being electron, muon or tau. If a neutrino interacts at a later time, then it now has a chance of acting like a different neutrino to the one it started life as – it is said to have oscillated to a different type. An electron neutrino begins life 100% electron-like, but after some time of the pedometers turning it will have a possibility of interacting as a muon neutrino. If it did, then it would produce a muon when exchanging a W boson and not an electron as would be naively expected from the 100% electron beginning.

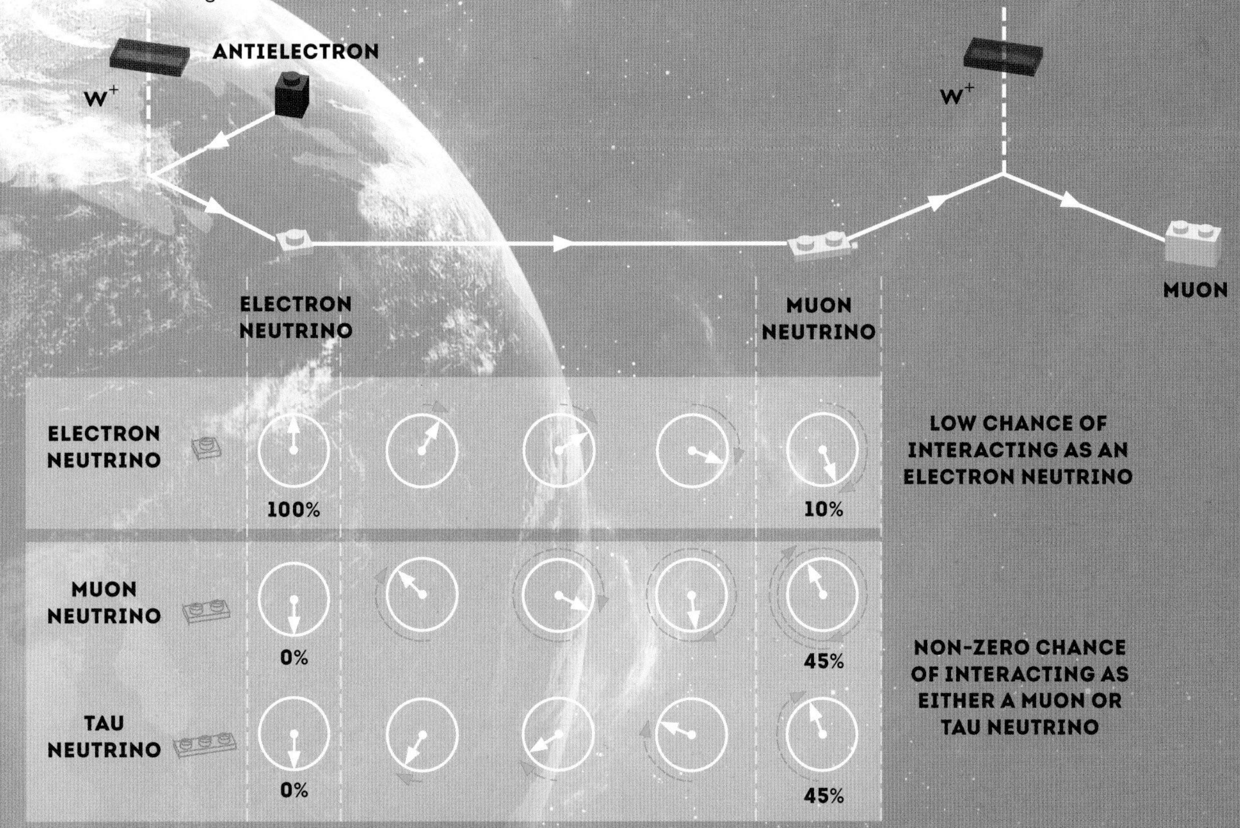

Neutrino oscillation is a purely quantum effect and occurs with neutrinos because they rarely interact with the world around them, allowing more time for their internal pedometers to spin and become ever more out of synch with one another. This explains the unexpected results of experiments looking at solar neutrinos and muon neutrinos from cosmic rays – they counted fewer neutrinos than expected because some of them had oscillated to other types which were not being recorded. The SNO detector, with its added ability to directly detect Z boson exchanges, was sensitive to all types of neutrino. SNO counted the number of Z^0 neutrino interactions expected to be the exact number predicted from the Sun.

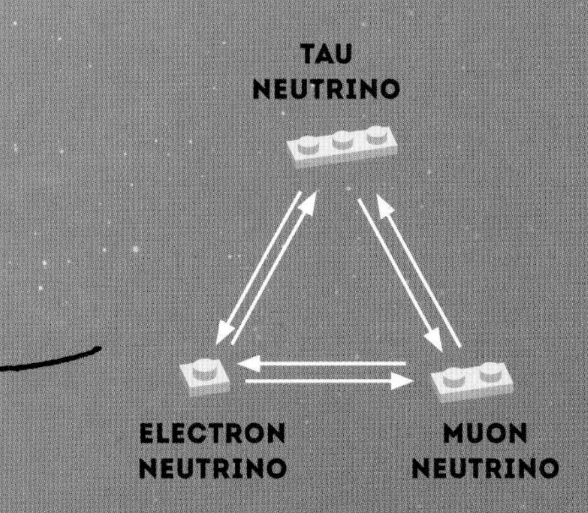

Predicted heavy neutrinos could explain why other neutrinos are so much lighter than any other particle

UNBALANCED SEE-SAW

The observed neutrino oscillation is only possible if neutrinos have a mass. Pedometers of particles slow down as they approach the speed of light, an effect of Einstein's relativity, and pedometers of particles travelling at the speed of light do not turn at all. To these massless particles, like the photon, a second is stretched out by relativity to infinity – time is effectively frozen. Their pedometers do not turn, so they can never get out of synch with one another, remaining at the same setting forever, meaning that something like neutrino oscillation could not happen. Because the pedometers of neutrinos do turn, they must be travelling slower than the speed of light, and so must have a mass.

PHOTON

Neutrinos are a million times lighter than the next lightest particle, the electron. Because of this huge difference, Standard Model neutrinos would have to have a very special relationship with the Higgs boson that would lead to them being so light. The required tinkering of the mathematics for such a relationship does not sit well with a lot of physicists, who feel that it undermines what is an otherwise elegant theory. A more popular method for the origin of neutrino mass does not involve the Higgs boson at all.

Neutrinos cannot interact with other particles through the electromagnetic or strong force. Only left-handed neutrinos and right-handed antineutrinos can interact with other particles through the weak force. This leaves right-handed neutrinos and left-handed antineutrinos with no force to interact with, making them indistinguishable from each other, which led Italian physicist Ettore Majorana to suggest a symmetry between the two. This symmetry between neutrinos and antineutrinos means that these neutrinos can act as their own antiparticle – something not possible for Standard Model neutrinos.

ETTORE MAJORANA

Ettore Majorana withdrew all his money on his way to Palermo in 1938 where on March 25th he boarded a boat bound for Naples, never to be seen again. There are many theories surrounding his disappearance, whichever may be correct, a true genius was lost that day.

MAJORANA NEUTRINOS

This splits the four types of neutrino grouped together in the Standard Model, left-handed and right-handed neutrino and antineutrino, into two distinct groups of Majorana neutrinos, each with different mass.

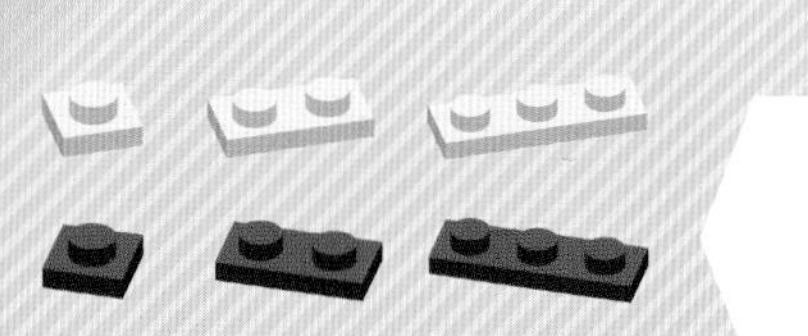

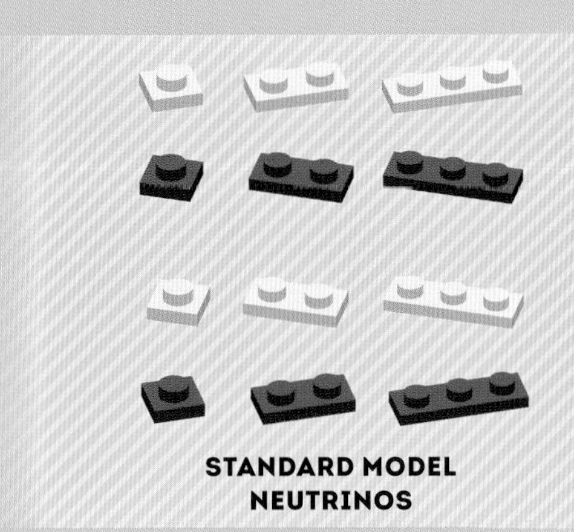

If one of these two new groups of neutrino were very massive they would, like an adult opposite a child on a see-saw, make the other neutrino group seem very light. This see-saw mechanism offers an explanation as to why the left-handed neutrinos and right-handed antineutrinos that we see interacting via the weak force seem so light. They only appear light because they are held up by a far heavier group of right-handed neutrinos and left-handed antineutrinos. This is the most popular theoretical explanation as to why neutrinos are a million times lighter than the electron. But this theory requires the existence of Majorana's version of the neutrino, and not the Standard Model ones of Paul Dirac.

If Majorana neutrinos exist, radioactive decay in which neutrinos are not emitted can also exist

TESTING THE BEGINNING OF TIME

To test whether Majorana neutrinos exist, we need to investigate the possibility of neutrinos acting as their own antiparticle. This is done by searching for a process known as neutrinoless double-beta decay.

Double-beta decay is the simultaneous beta decays of two neutrons inside an atomic nucleus, emitting two electrons and two electron antineutrinos. It is found to exist in a handful of radioactive isotopes.

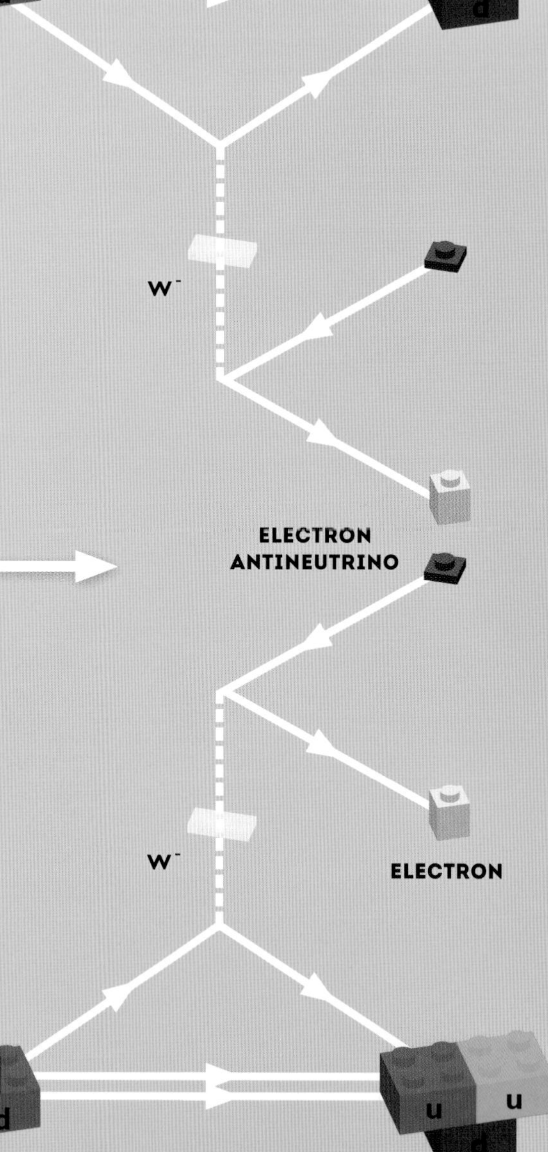

In standard double beta decay some of the energy released is taken away by the neutrinos, giving the electrons a wide range of possible energies. In neutrinoless double-beta decay, however, there are no neutrinos to take energy away and so the ejected electrons would carry away the maximum possible energy. To search for neutrinoless double-beta decay, experiments carefully measure the energy of electrons. If there is a significant excess of electrons with highest possible energy then it suggests they are coming from neutrinoless double-beta decay, not regular double-beta decay. So far they have drawn a blank, but the search for Majorana neutrinos continues.

The right-handed Majorana neutrinos, and left-handed antineutrinos, would have to be very massive indeed to push the mass of observed neutrinos as low as it seems to be. The simplest version of the see-saw mechanism pitches their mass at around 10^{15} GeV, much higher than the Large Hadron Collider, or indeed any future planned particle accelerator, could reach. Such massive particles could only exist virtually today, blinking out of existence in far less time than a W boson. This makes interactions between them, like annihilation, extremely rare. Measuring the chance of neutrinoless double-beta decay occurring is a direct measurement of the lifetime and therefore the mass of these heavy neutrinos.

Heavy Majorana neutrinos could only have existed as real particles when the Universe was very young and hot, a tiny fraction of a second old. At these energies it is thought that not only the electromagnetic and weak forces merge into one electroweak force, but that the symmetry describing the strong force may also combine with this electroweak force into one all-encompassing force. So understanding whether neutrinos are Majorana or not is asking much grander questions about our very young Universe and the origin of the forces.

The neutrinoless version is exactly as you would expect, two simultaneous beta decays but no neutrinos released. This is only possible if neutrinos are Majorana in behaviour (their own antiparticle) and annihilate one another.

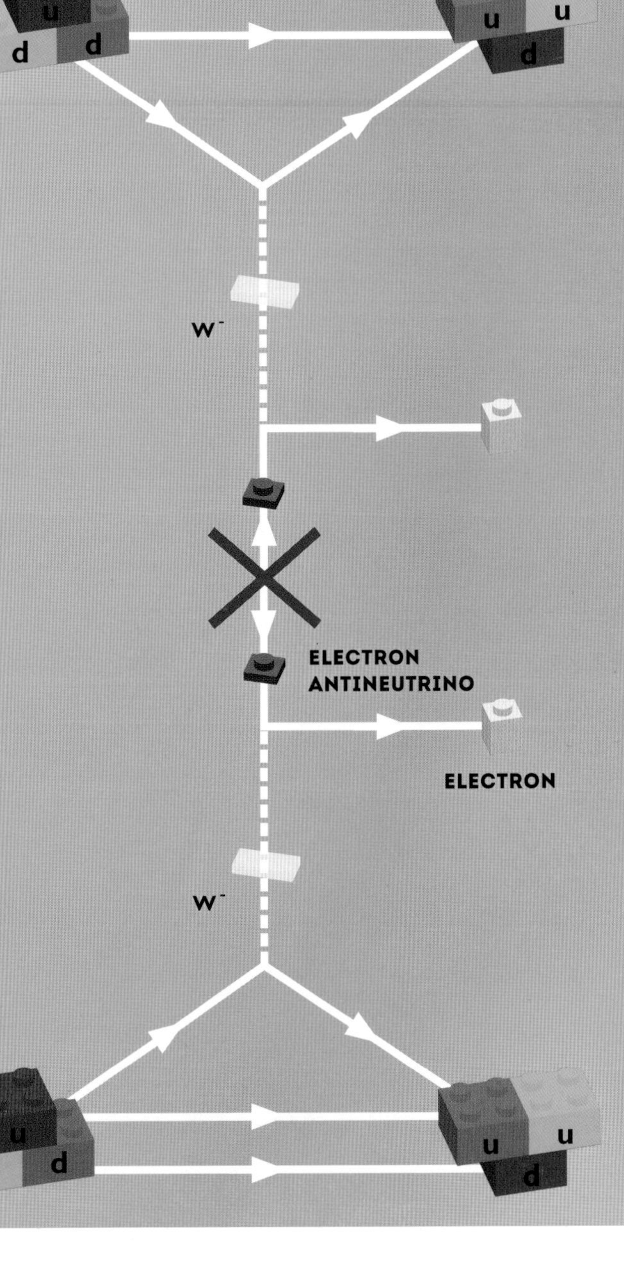

BETA DECAY

Beta decay is a form of radioactive decay in which a neutron within an atomic nucleus decays, forming a proton, electron and electron antineutrino.

Legitimate antiparticles can only be created by flipping both charge and chirality

A SUPREME SYMMETRY

The bias of the weak force, preferring just left-handed particles and right-handed antiparticles, renders some symmetries no longer symmetric. So what is the true symmetry between particles and antiparticles which seems to be played out whenever particle and antiparticle annihilate one another or a boson creates a particle–antiparticle pair?

CHARGE INVERSION

In chapter 1, I mentioned that to change a particle into an antiparticle we only had to invert/mirror its electric and strong (colour) charge. This changes the charge of each particle, but doesn't affect other properties such as whether a particle is left- or right-handed. So using a charge inversion we could change a left-handed particle to a left-handed antiparticle. However, as only right-handed antiparticles interact with the weak force, we would end up with a particle that does not interact with the weak force any longer. As the particle's interactions have changed, this would not be a symmetry.

PARITY INVERSION

In order to change whether a particle is left- or right-handed, we need another mirror. This mirroring is called a parity inversion and flips all of space on its head, not just left and right but up and down and front and back as well. However, parity inversion does not change the charge of a particle. Just changing left-handed to right-handed particles is not a symmetry as you still end up with right-handed particles which do interact with the weak force.

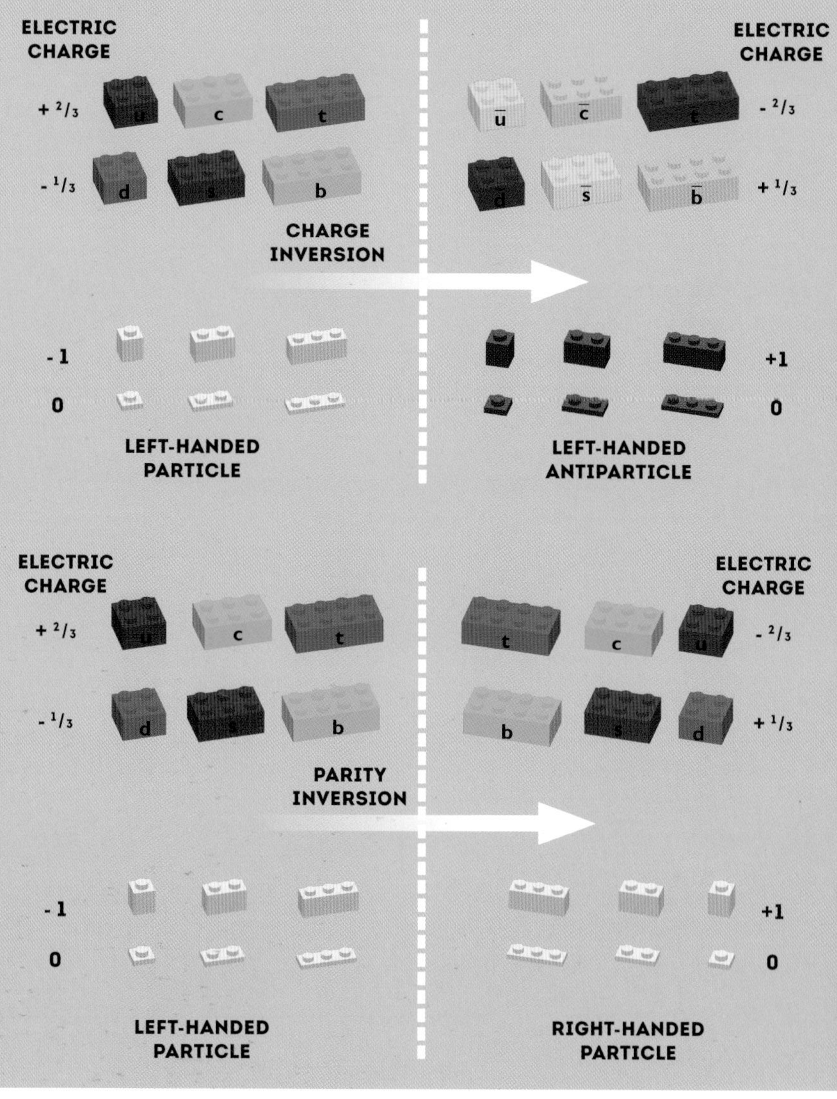

CHARGE AND PARITY INVERSION

Combining both charge and parity inversions changes every left-handed particle to a right-handed antiparticle. This leaves all particles with the same size charge and the same ability to interact with the weak force. It is a true symmetry between particles and antiparticles. This combined charge–parity symmetry is seen in cosmic rays and particle accelerators today, every time a force-carrying boson produces a particle–antiparticle pair or when particle–antiparticle pairs annihilate one another to form a boson. Each interaction seems balanced with the same number of particles and antiparticles.

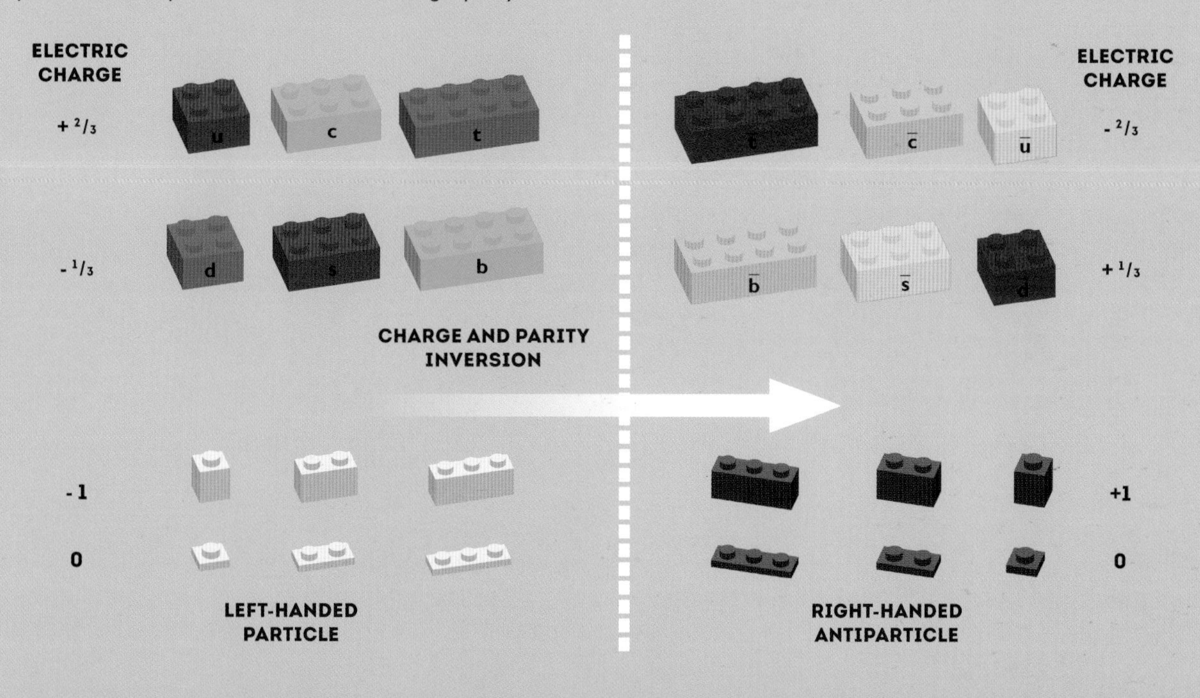

PARITY VIOLATION

If parity were a true symmetry of Nature, there should be no interaction that shows a preference for a particular direction in space. All electromagnetic and strong force interactions certainly didn't care for any particular direction – emitted particles distributed evenly. In 1956 the Chinese theorists Tsung-dao Lee and Chen-Ning Yang pointed out that parity symmetry had yet to be tested in weak force interactions. Lee persuaded his colleague at Columbia University, Chien-Shiung Wu, to investigate their hunch that parity was not a symmetry respected by the weak force. Wu used cryogenically cooled cobalt-60 atoms all aligned to point in a single direction by a strong magnetic field. She carefully measured the direction of the electrons emitted when the cobalt-60 decayed through beta decay. If they respected the symmetry of parity then they would be emitted evenly in all directions. Instead what was observed was that most electrons were emitted in just one direction, despite photons emitted alongside them during the decay being emitted evenly in all directions. This showed that the particles coming from the weak force beta decay had a preference for one direction in space, which could only be possible if the weak force violated parity. We know now that this is because it interacts only with left-handed particles and right-handed antiparticles, while the electromagnetic and strong forces do not care which they interact with.

Nature is not symmetric, it treats particles and antiparticles differently

A CRACK IN THE SYMMETRY

The visible Universe today is made from the twelve matter particles of the Standard Model, aside from special locations where energy runs high enough to create equal amount of both particles and antiparticles. In Chapter 2, I skipped over discussing where all of these matter particles came from, jumping straight from the pure-energy Big Bang to a soup of quarks and electrons ready to form atoms. Now is the time to talk.

THE BIG BANG

10^{-36} SECONDS
10^{28} K
10^{19} GeV

If charge–parity symmetry (discussed on the previous pages) were a true symmetry, then from the Big Bang's pure boson energy, particles and antiparticles should have been created in equal numbers. As the Universe cooled, the particles and antiparticles would no longer be able to resist each other's electric attraction, eventually meeting to annihilate one another to form photons. The cold Universe today should be nothing more than a bath of light from particle–antiparticle annihilations and weakly interacting neutrinos. Yet here we are.

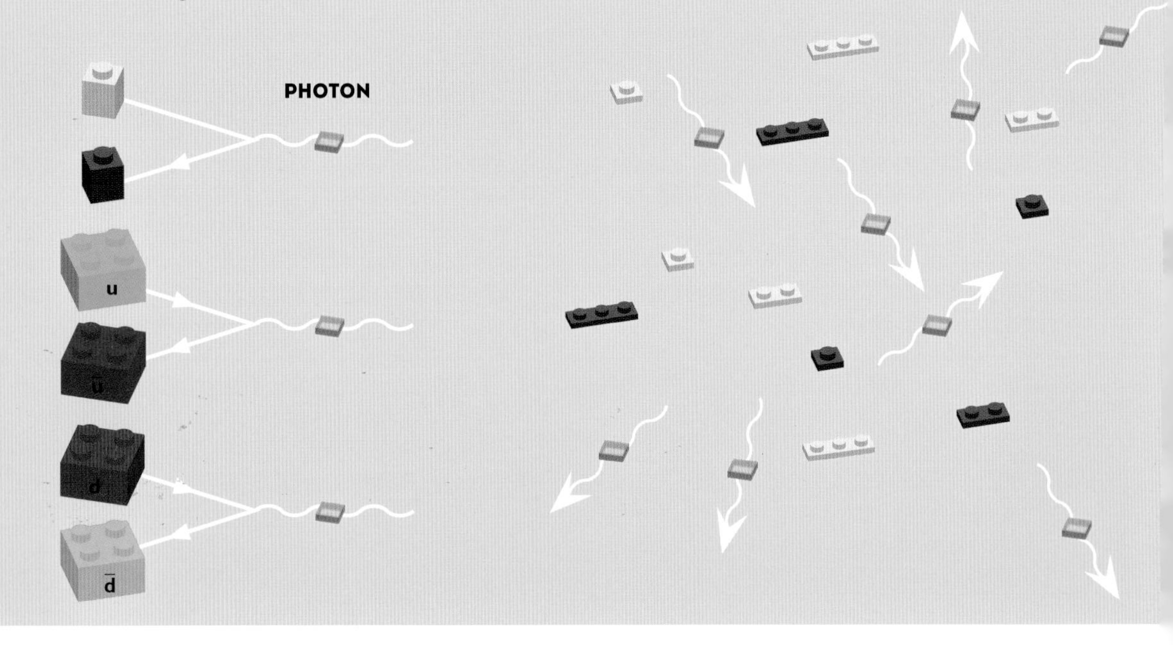

It is in the mixing of quarks (pages 128–9) and neutrinos (pages 146–7) that the answer is thought to lie. While the change in charge–parity symmetry leaves particles with the same force interactions overall, it introduces a small change to their internal pedometers. This can be thought of as a small defect which affects the speed at which the pedometers tick. The defect has an opposite effect for particle and antiparticle. If the defect sped up a certain particle pedometer, then it would slow down the parity–charge mirrored version pedometer of an antiparticle. This defect is known as charge–parity violation (CP-violation) and creates a very tiny crack in an otherwise perfect symmetry. This crack meant that when all of the antimatter had been annihilated through contact with matter, a small residue of matter was left behind, rather like tea leaves in the bottom of a cup.

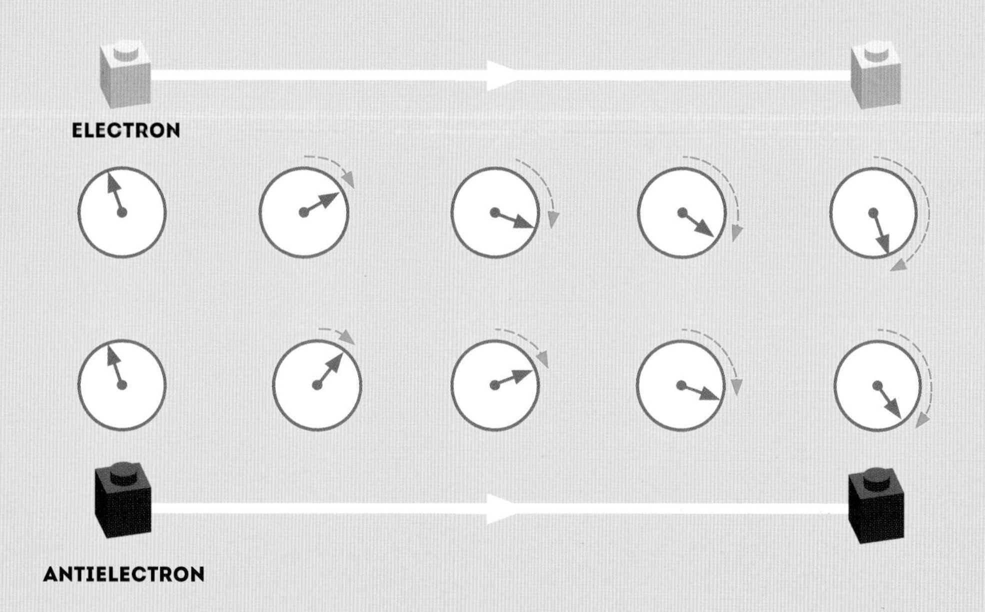

Day to day we see this symmetry as perfect, with equal amounts of matter and antimatter – like the charge (colour) and parity (flipping of left to right and up to down) of this M C Escher-inspired tessellation made from white particles and black antiparticles. On closer inspection, however, we see that while on the face of it the original and final pictures look the same, there is a subtle difference. Spotted it yet? It is in the eyes of the fish. The subtle difference in internal pedometers manifests itself as a real-world bias of less than 1 part in 10,000,000 with Nature preferring matter over antimatter in rare interactions.

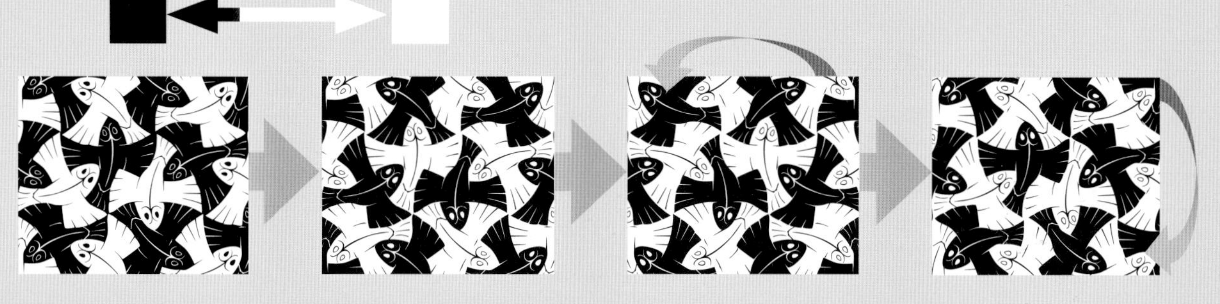

Imbalance between quarks and antiquarks has been found from searching trillions of interactions

A NEEDLE IN A HAYSTACK

Processes which show a preference for matter over antimatter are very rare which means large amounts of data need to be sifted through to find evidence of them. The pedometer defect which arises in quark mixing is measured by looking at millions upon millions of heavy meson decays. The first evidence for charge-parity violation came from looking at the decay of neutral kaons.

There are two distinguishable versions of a neutral kaon: the K^0 containing a strange antiquark and down quark, and the $\overline{K^0}$ containing a strange quark and down antiquark. Just like the mixing of neutrinos (see pages 146–7), mixing of quarks leads to these neutral particles oscillating between these two possible quark configurations.

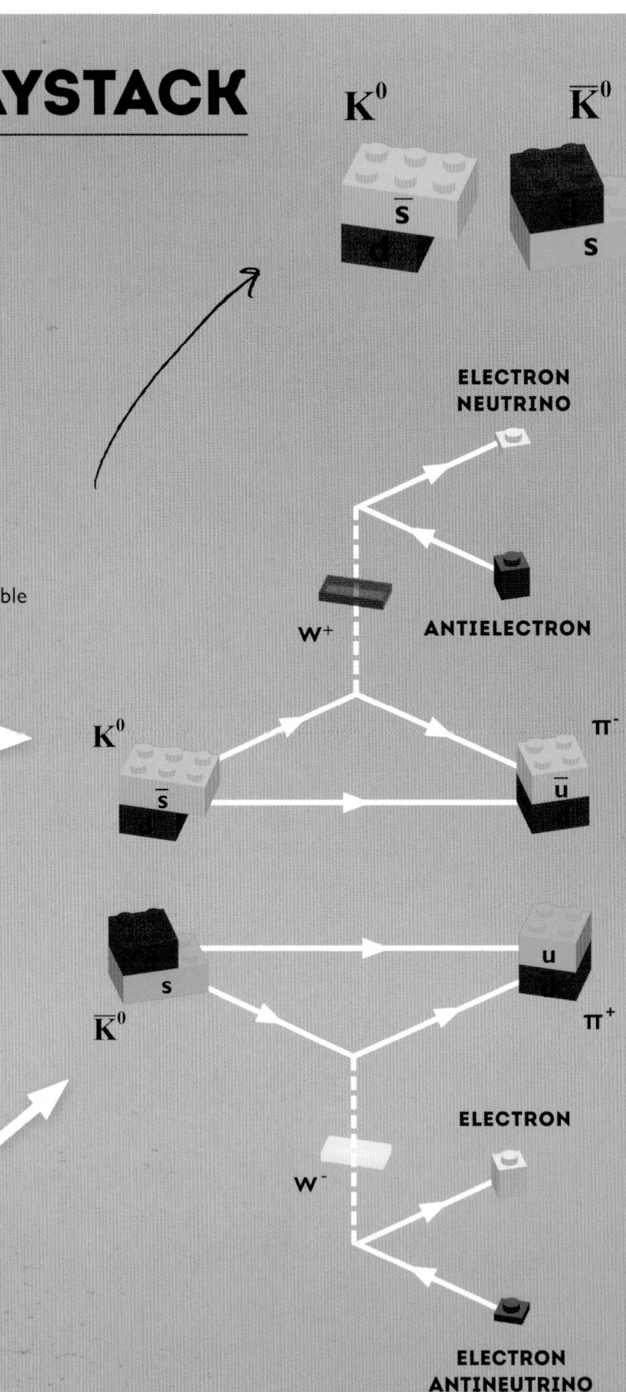

It is always the strange part of the particle that decays, thanks to the weak force, as it is the heavier quark or antiquark. A strange antiquark, inside of every K^0, will decay into an up antiquark by exchanging a W^+ boson which will produce an antielectron.

A strange quark, inside of a $\overline{K^0}$, decays to an up quark by exchanging a W^- that goes on to produce an electron.

If matter and antimatter are treated equally, then there should be an even 50–50 split in how often each kaon decay takes place, but there isn't. The symmetry is violated by the defect in the internal pedometers of quarks, changing the probability of K^0 and $\overline{K}^0$ oscillation in favour of the K^0 electron (matter) producing decay. The difference is tiny – only 0.3%. Given the number of kaons which might have decayed in the lifetime of the Universe, this alone cannot account for the entire matter of the Universe around us.

To obtain the full picture of CP-violation among quarks, experiments have also looked at the decay of heavier bottom and charm quark containing mesons. Experiments such as Belle in Japan and the latest LHCb experiment part of the LHC in CERN sift through millions of meson decays searching for a bias. Here too they have found that bottom and charm-containing quarks prefer matter to antimatter in their decay. Once all is totalled up however, there is still not enough violation of charge–parity among the quarks to explain how all this matter came to exist.

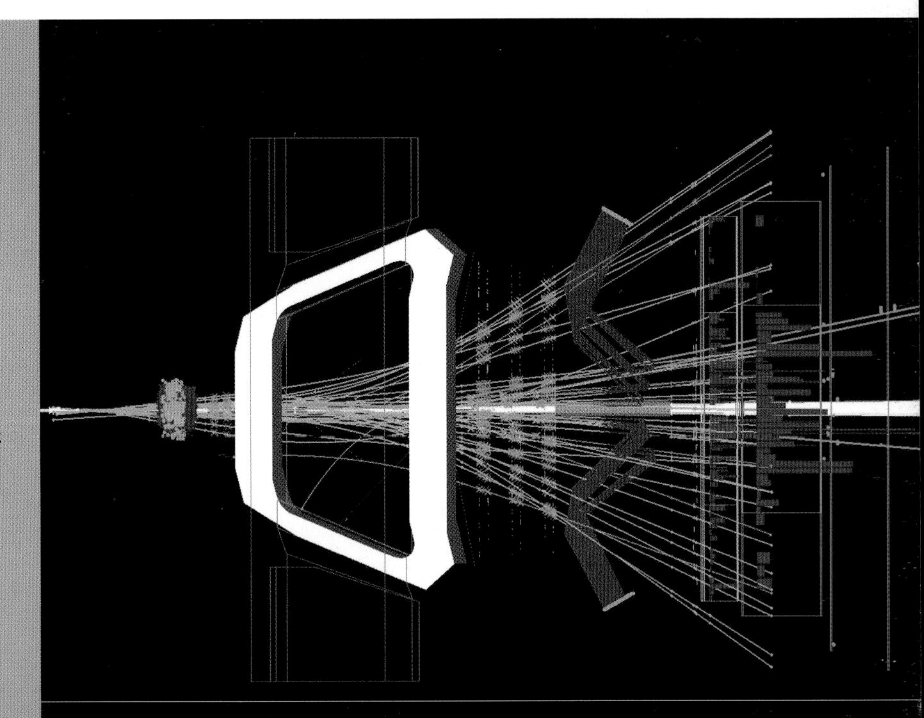

Particle tracks proton-proton collisions seen by the LHCb (Large Hadron Collider beauty) detector.

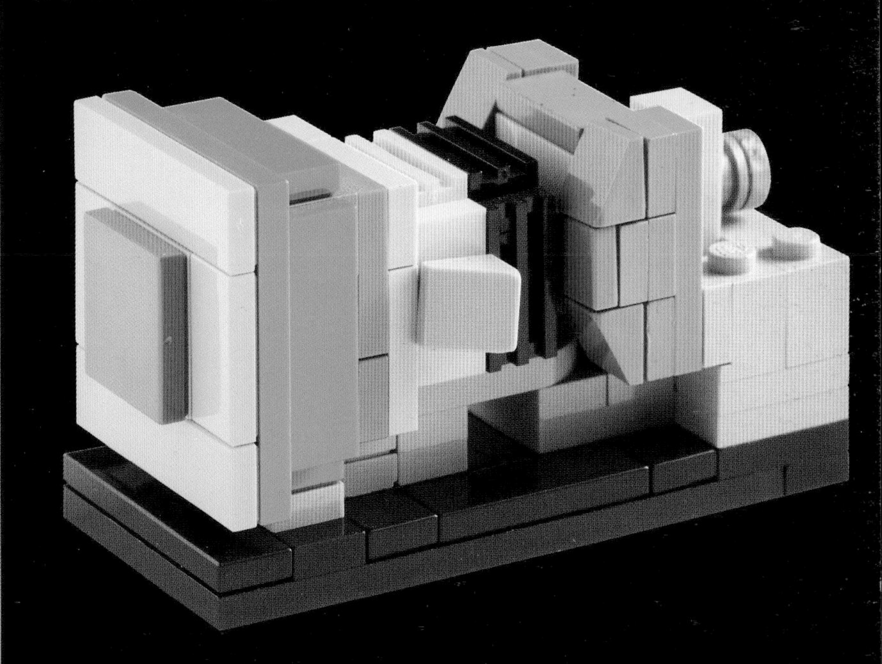

A plastic model of the LHCb detector showing the magnet region in blue, Cherenkov detectors in yellow and muon detectors in green. Designed by Nathan Readioff.

KAONS

Kaons were the second type of meson particle to be discovered. They are strange particles as they contain one strange quark or antiquark.

Scientists are looking to leptons to provide evidence for charge–parity violation

LOOKING TO LEPTONS

Now all eyes are on the leptons, neutrinos in particular. Mixing between leptons allows neutrinos to oscillate and, just like the neutral kaons, any difference in the probabilities of how neutrinos or antineutrinos oscillate would be a direct result of a crack in symmetry between particles and antiparticles.

Determining the probabilities of kaon oscillations was relatively easy compared with neutrinos. The kaons have a finite lifetime before decaying into measurable charged particles and could be produced at known locations in accelerators, where two beams of particles collide head on.

Neutrinos on the other hand interact very rarely with the world around them and have no pressure encouraging them to do so. Also, while beams of neutrinos can be formed, there is still no way of predicting exactly where in a detector a neutrino might interact.

These factors combine to make understanding neutrino oscillations a very different game altogether. Only now in the 21st century, using man-made beams of neutrinos, are we beginning to truly understand the likelihood of neutrinos oscillating between different types. While we have measured how the internal pedometers of neutrinos turn, there is still uncertainty surrounding their oscillation behaviour. But we are close to getting answers.

The current generation of neutrino oscillation experiments Tokai to Kamioka (T2K) based in Japan and the US-based NOvA are now honing in precisely on this information. Once there is an accurate understanding of how neutrinos oscillate, it will be a case of testing if antineutrinos behave in exactly the same way. To do this, experiments change the magnetic field in their beam to focus the opposite electrical charge of pion, so that antineutrinos are produced from the decay of negative pions. Hints of charge–parity violation are already showing up in these early comparisons of neutrino and antineutrino oscillation, but more evidence is needed.

HOW TO MAKE A NEUTRINO PARTICLE BEAM

PROTON

1

2

1 Take high-energy protons from a particle accelerator.

2 Smash the protons into a target made from a light chemical element, e.g. carbon, which creates a shower of pions.

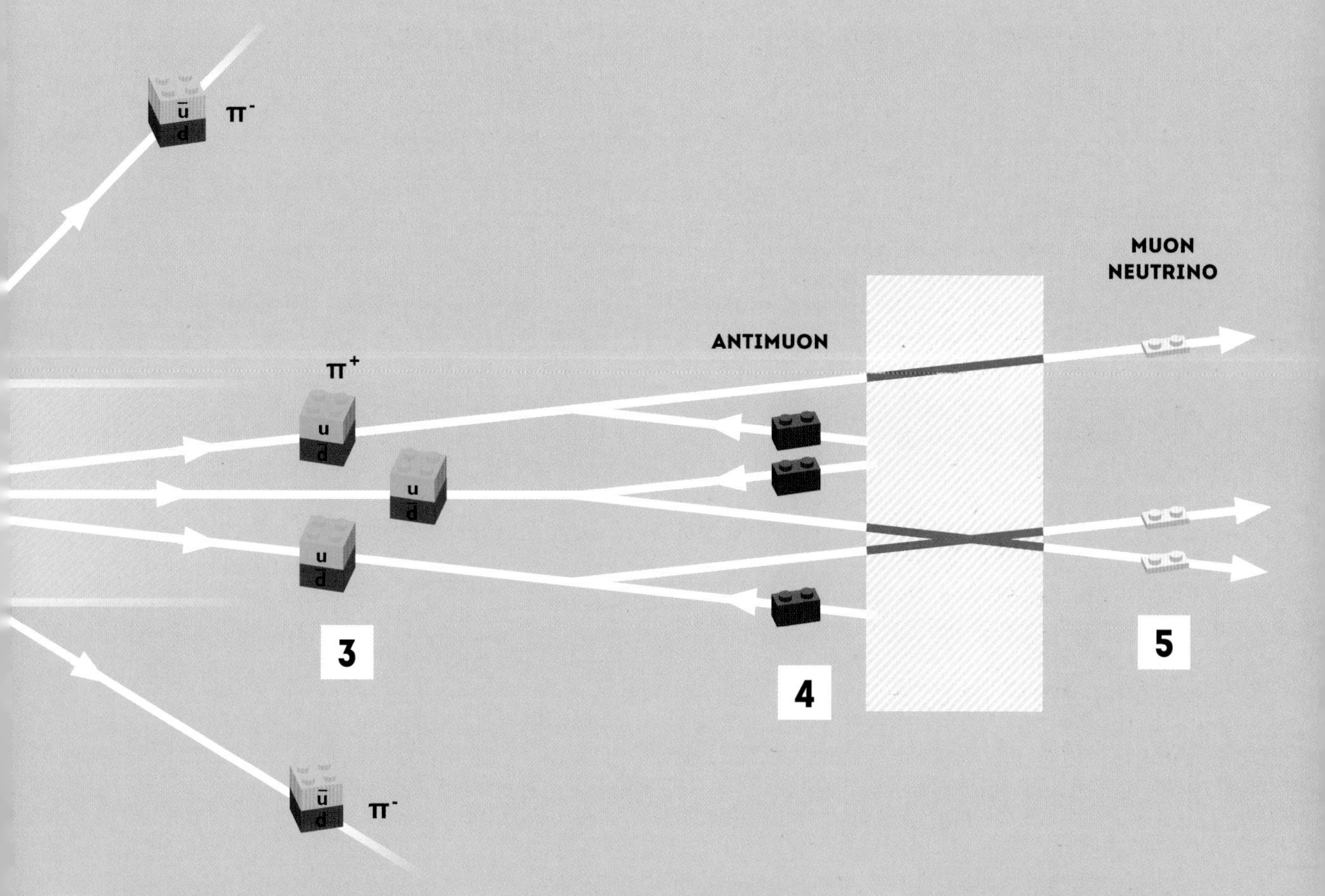

3 A powerful magnetic field is used to bend one electric charge of pions into a focused beam, while at the same time bending the other electric charge of pions in the opposite direction away from the central beam.

4 Allow the pions time to decay into neutrino–muon particle–antiparticle pairs.

5 Stop the muons with a big block of concrete or iron. Only the neutrinos make it through as a neutrino beam to be detected kilometres away.

LEPTONS

Leptons make up one half of the fermion matter-building particles. They do not feel the strong force and come in two types, electrically charged (electron, muon and tau) and electrically neutral neutrinos (electron neutrino, muon neutrino and tau neutrino).

With more leptons than antileptons, leptons must be able to change to baryons for balance

GRAND IDEAS

Charge–parity violation, the bias for particles over antiparticles, has been measured for quarks but is smaller than the amount needed to explain our matter-dominated Universe. The bias for particles over antiparticles must, therefore, be larger among the leptons to make up for this shortfall. This would mean that charge–parity violation on its own would not be enough to produce all of the quark-containing baryons in the Universe today, only the leptons.

You may have noticed that there are other symmetries of Nature which can be considered conserved in particle interactions. One conserved quantity is that of baryon number: one third of the total number of quarks minus the total number of antiquarks.

As leptons are not made of quarks at all, they have a baryon number of 0, but they do have their very own conserved lepton number. This lepton number is +1 for all charged leptons (electron, muon and tau) and their neutrinos, and -1 for the antiparticle versions.

Look back through the book and you will see that every interaction drawn conserves these baryon and lepton numbers, for example beta decay.

BARYON NUMBER, B = $^1/_3$ [NUMBER OF QUARKS - NUMBER OF ANTIQUARKS]

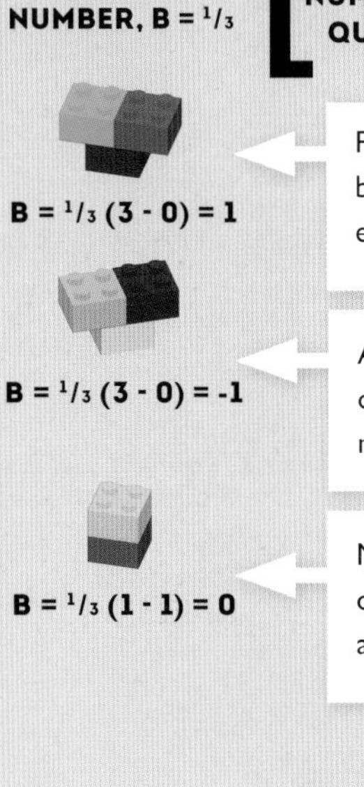

Protons, neutrons and other baryons have a baryon number equal to 1.

Antiprotons, antineutrons and other antibaryons have a baryon number of -1.

Mesons, with an equal amount of quarks and antiquarks, have a baryon number of zero.

LEPTON NUMBER, L = [NUMBER OF LEPTONS - NUMBER OF ANTILEPTONS]

CHARGE-PARITY VIOLATION

The symmetry which exists between particles and antiparticles relates to a transformation in the charge of a particle and a mirroring of space. It is thought that a violation of this symmetry led to a preference in particles over antiparticles in the early Universe.

ELECTRON ANTINEUTRINO

L = -1

L = 1

B = 1

W⁻

ELECTRON

B = 1

d d u

u u d

L = 0

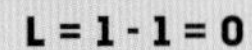

If baryon number and lepton number are each individually conserved, then a baryon cannot become a lepton in any of these interactions. With this in mind, the greater expected bias for leptons over antileptons compared to the bias for quarks over antiquarks, would lead to many more leptons than baryons, but this is not seen today. There must be some process where leptons can change into baryons to address this imbalance and for that, there must be a violation of baryon and lepton number.

At extremely high energies, such as those immediately after the Big Bang, it is thought that the symmetries explaining the strong, weak and electromagnetic forces combine. The new symmetry that emerges does not care if a particle is a baryon or lepton, only that the total number of both are conserved; the sum of baryon number minus lepton number (B-L).

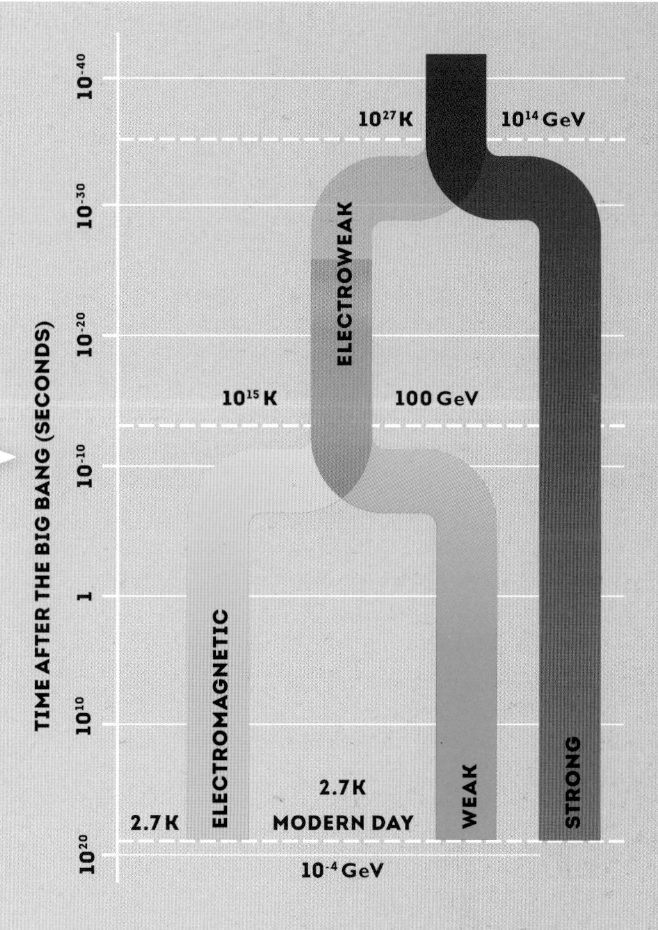

SPHALERONS

New processes involving new entities called sphalerons allow leptons to become baryons, increasing the baryon number but decreasing the lepton number by the same amount. The overall electric charge entering and exiting a sphaleron process is zero and there is no net colour change, as the symmetry describing the sphaleron must still respect the other symmetries in the Standard Model.

Sphalerons are not particles, but a modification to the particles' internal pedometer which replaces leptons with baryons until they become balanced. This is why we can't draw a Feynman diagram to show the changes. With sphalerons and charge–parity violation, we have an explanation for where all of the building blocks of matter first came from in the Standard Model.

Right now there is no evidence for violation of either baryon or lepton number and so this is all just a theory, but the best one we have. Experiments like Super Kamiokande in Japan sit waiting to see if a proton will decay into a neutral pion and antielectron, a process only possible if baryon number and lepton number are violated but the new grand unified symmetry of B-L still holds.

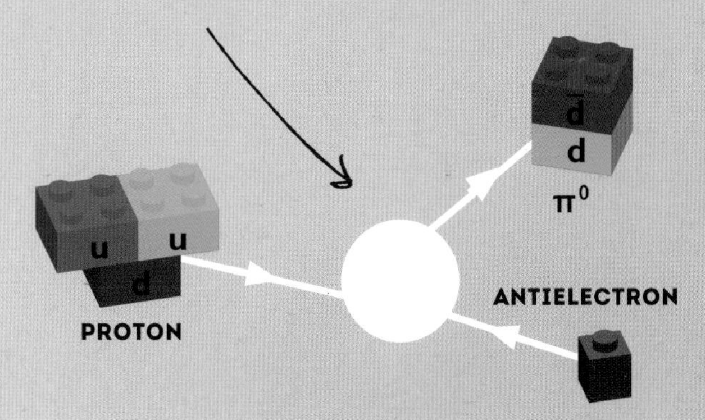

To solve the big questions, we need a new theory that works at high energies

1.22×10^{19}

PROBLEMS

The quantum theory of electromagnetism, quantum electrodynamics, did not work when first devised. As you look closer and closer at Feynman diagrams of any particle interaction, the uncertainty principle allows for loops of particle–antiparticle pairs (see pages 116–7). Two electrons can scatter off one another, but between incoming and outgoing there can be 3, 21, 100,986 or indeed any number of virtual particles, born for a fleeting moment from borrowed energy. This gave nonsense infinite possibilities.

To get sensible answers from calculations, theorists had to place an upper energy cut-off beyond which the Standard Model does not work. This energy ceiling is called the Planck scale and states that no Standard Model particle can have energy or mass greater than 1.22×10^{19} GeV. This removed the possibility for infinite virtual quantum loops and returned sensible (normal) answers. This is an obvious sign that there has to be something beyond the Standard Model, beyond this artificial energy cut-off. The Standard Model was a low-energy approximation to some grander theory – like Newton's mechanics to Einstein's relativity.

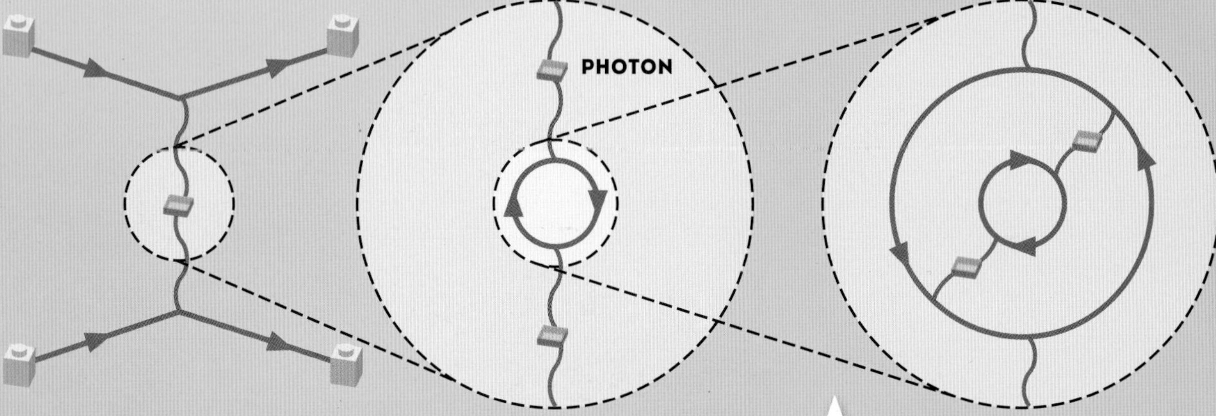

PLANCK SCALE

The Planck scale is the upper limit in energy at which the Standard Model works, beyond this high energy a new theory is needed.

When scrutinising particle interactions in ever finer detail, things become more complicated. What seems to be a simple exchange of a photon might also include loops of particles and antiparticles or a number of additional photons.

PLANCK SCALE *NEW* THEORY NEEDED

STANDARD MODEL

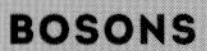

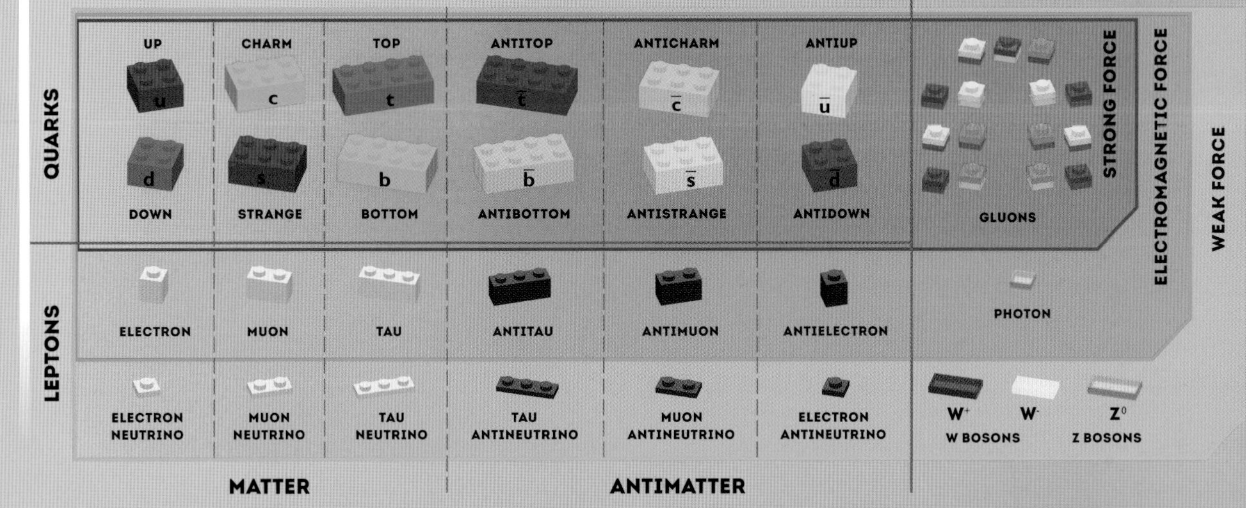

Even with this cut-off, the Higgs boson should not be as light as it has been measured. The loops of fermions that appear in any Higgs interaction should increase the measured mass of the Higgs boson up as far as the Planck mass – which is around the same size as a flea's egg, not much by human standards but huge for a single fundamental particle! This is a big argument against the Standard Model and is known as the hierarchy problem.

H

The Standard Model, like all quantum theories is only able to tell us how things happen. All of quantum physics is a framework to explain what we see in experiments. It cannot tell us *why* we get the results that we do. This is a big issue for many physicists, as it is like politicians informing the public how a big decision is made in government but not why. The 'why' is the deepest understanding of all and the driving force behind science to find theories beyond the Standard Model.

The Standard Model does not provide any candidates for dark matter or dark energy

DARK SECTOR

The Standard Model describes the entire visible Universe. However, measurement of the Universe at the largest cosmological scales tells us that this is just 5% of the total energy out there. The rest is dark, not visible through the electromagnetic force. The largest share, about 69%, is a dark energy which is working against gravity to expand our Universe at an increasing rate. Empty space, vacuum, is actually teeming with activity on the smallest scales as virtual particle–antiparticle pairs pop in and out of existence. Although this could be a source of dark energy, it could not account for even the tiniest fraction, in fact just $1/10^{120}$. We are still very much in the dark about dark energy.

The Alpha Magnetic Spectrometer (AMS) attached to the outside of the International Space Station is searching for possible decays of dark matter.

Around 25% of our Universe's energy seems locked up in dark matter, which causes gas and stars to clump together under gravity to form galaxies. The Standard Model does not provide suitable candidates for what particles dark matter is made from. Although neutrinos fit the bill, they are far too light and therefore travel far too fast to clump together through gravity – fast-moving neutrinos would smear galaxies out more thinly than we see them today.

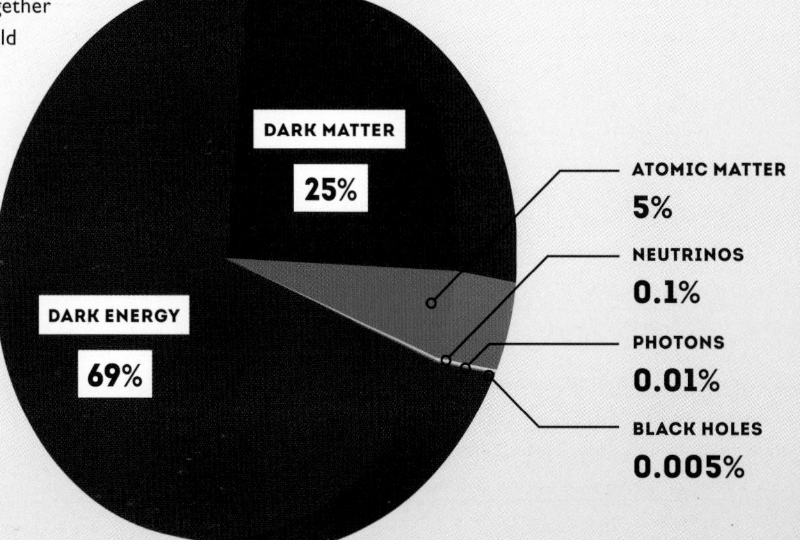

DETECTING WIMPS

Dark matter is thought to be made up, like normal matter, of particles. It would only be possible for these dark matter particles to interact via the weak force or gravity – otherwise they would be easily detectable – and have been given the name of WIMPs (Weakly Interacting Massive Particles). There are three ways of possibly detecting a WIMP:

1 Using the world's most sensitive particles detectors, look for a direct hit from a dark matter (DM) particle imparting energy to a detectable Standard Model (SM) particle. This is being done in deep caverns around the world to ensure the detectors are shielded from cosmic rays and other particles which might fake a dark matter detection. There have been a number of false alarms among these experiments, but evidence is beginning to mount. It will be a few years yet before it is clear whether dark matter is indeed the cause of these results or if it is a statistical blip on the radar screen.

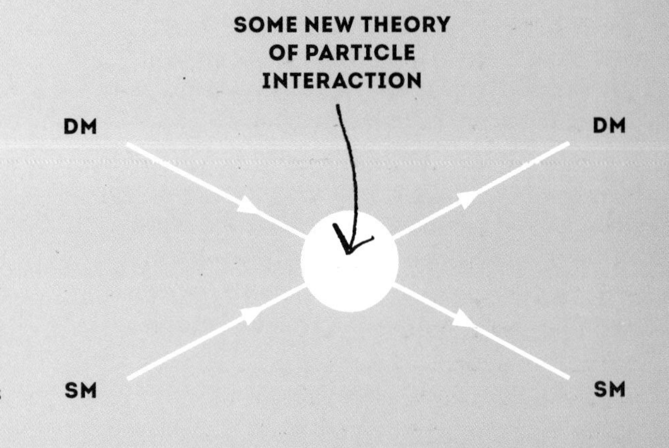

DM
SM
DM
SM

2 Look for evidence of dark matter particles annihilating one another, forming Standard Model particles which can be measured. The Alpha Magnetic Spectrometer (AMS-02) on board the International Space Station carefully measures the energy of particles in cosmic ray collisions for hints of dark matter interactions. The Fermi Gamma-Ray Telescope is also looking for high-energy photons which could only have come from dark matter particles annihilating one another.

3 Smash Standard Model particles into each other and hope that they will produce dark matter particles. This search is being done at the ATLAS and CMS detectors at the Large Hadron Collider. The dark matter particles would, like neutrinos, leave the experiment unseen and so to search for dark matter you look for unexplained missing energy. So far there is no solid evidence for any such particles.

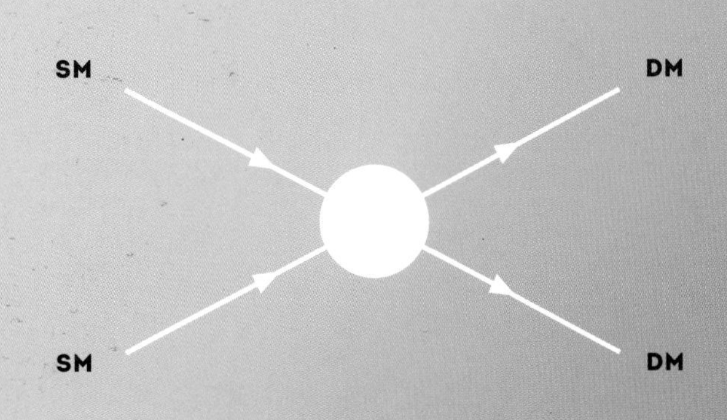

A new symmetry predicts new particles which may be the source of dark matter

SUPER-SYMMETRY

Sticking with the success of symmetries, one extension to the Standard Model suggests that there exists a symmetry between the fermion (matter-building particle) and the boson (force-carrying particle). Grandly called super-symmetry, it predicts a whole new set of super particles or sparticles, where the super-fermions behave like bosons and the super-bosons behave like fermions. It is able to solve a lot of the problems with the Standard Model.

The new symmetry requires more particles, including more than one Higgs boson – in fact in the simplest super-symmetry theory there are five! Three electrically neutral Higgs bosons, one of which is the one detected at the LHC, and two electrically charged Higgs bosons. The super-symmetric partners to these Higgs bosons are the Higgsinos (higgs-ee-no-s).

The names of the fermion super partners are not too imaginative, simply an addition of an s (for super) in front of their name: quarks become squarks (pronounced skwark not s-kwark) and leptons become sleptons. The electron becomes the selectron, muon the smuon, and tau the stau – the neutrinos become the sneutrinos. The gluons' super-partners are called gluinos.

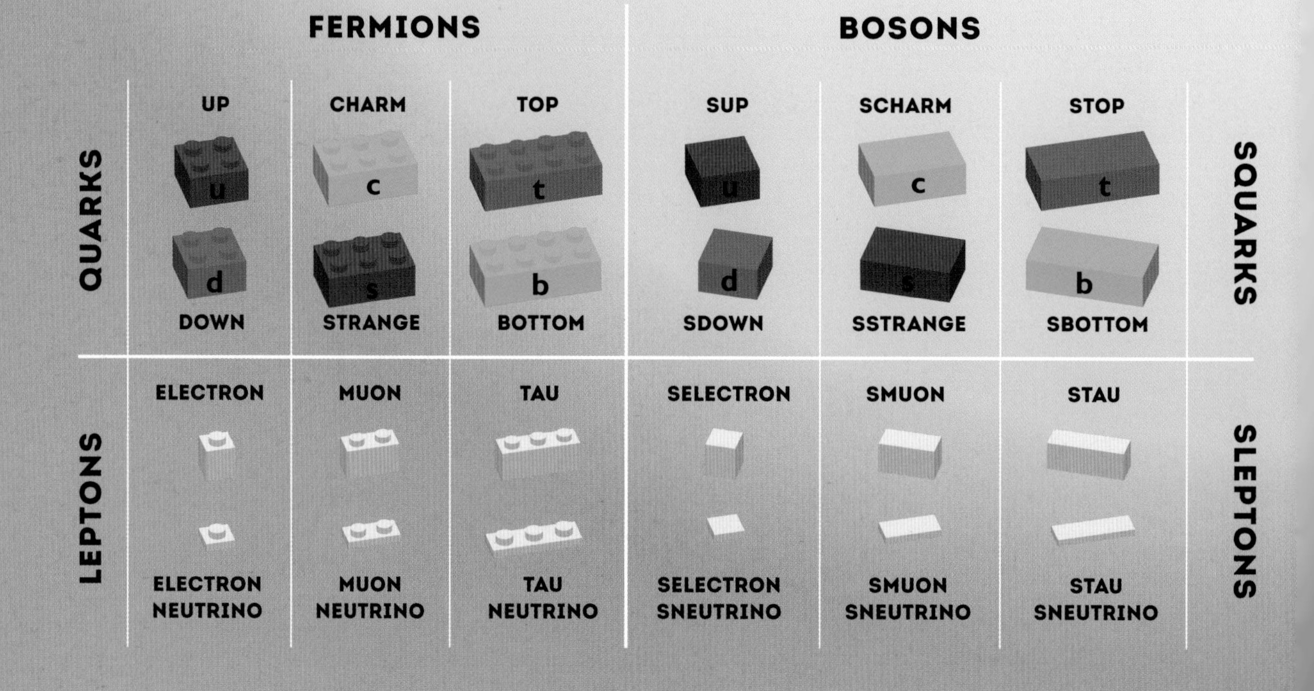

When it comes to the super-partners of the electroweak bosons things get complicated by extra Higgsinos. The result from the mixing together of the super-bosons by the super-Higgsinos creates four electrically charged sparticles called charginos ($\overline{\chi}_1^+, \overline{\chi}_1^-, \overline{\chi}_2^+, \overline{\chi}_2^-$) and four electrically neutral sparticles called neutralinos ($\overline{\chi}_1^0, \overline{\chi}_2^0, \overline{\chi}_3^0, \overline{\chi}_4^0$) – not to be confused with neutrinos!

The charginos can decay into neutralinos by exchanging a Standard Model W boson and neutralinos can decay into the lightest neutralino by exchanging Standard Model Z^0 bosons. The lightest neutralino $\overline{\chi}_1^0$ is thought to then be stable, unable to decay, which makes it the perfect candidate for dark matter WIMPs.

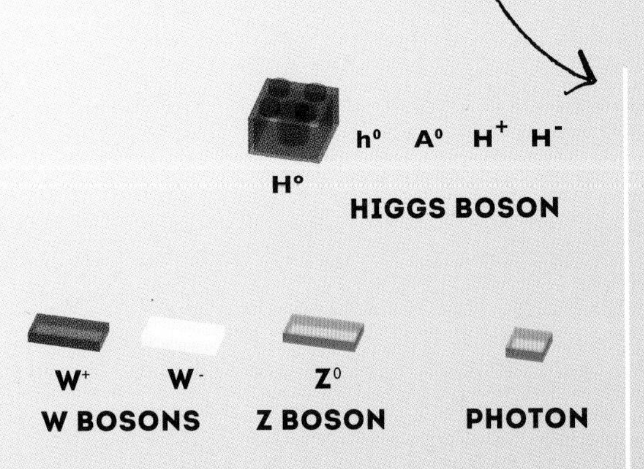

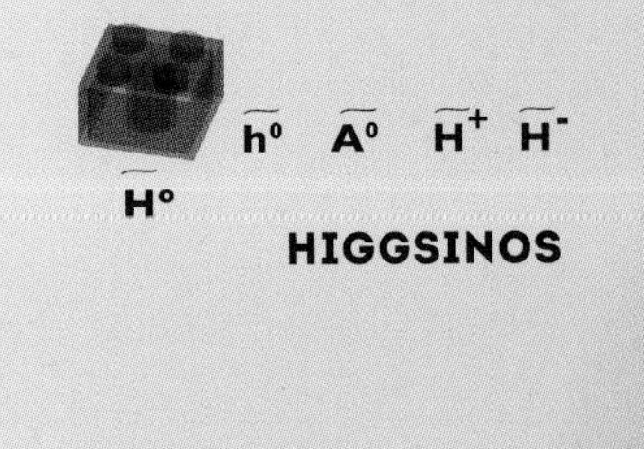

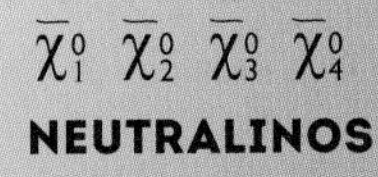

The hierarchy problem which results in the expected huge mass of the Higgs is solved by virtual sparticle loops (dashed lines in Feynman diagrams as sparticles are bosons too). The sparticle loops cancel out the particle loops which otherwise increase the measured mass of the Higgs boson in the Standard Model.

Super-symmetric particles, if they exist, should be around the same mass as the Higgs boson which is central to the theory. However, at the time of writing not a single super-symmetric particle has been observed at the LHC and data from the later runs show that no additional Higgs bosons have been seen, only the one required by the Standard Model. As data continues to grow, and at a higher energy of 13 TeV, so too does scepticism about super-symmetry. The next few years could prove to be make or break for this theory.

CLOCKWORK THEORY

Another theory which explains away the lighter-than-expected mass of the Higgs is clockwork theory. In this theory the Higgs is not directly linked to the Planck energy scale of gravity but instead its relationship has to pass through a number of intermediary interactions. This protects the Higgs from needing to have such a large mass as is otherwise expected from the Standard Model.

To go beyond the Standard Model we may have to embrace theories with extra dimensions

BUILDING WITHOUT BRICKS

Gravity is a force like no other. In Einstein's general relativity, our most tried and tested understanding of the force, it affects the space–time stage upon which particles perform. While the other three forces of Nature determine the set routines and choreography of their dance (right), gravity affects the very stage upon which particles perform by curving it (far right).

STRING THEORY

Super-symmetry unifies the strong, weak and electromagnetic forces, but something radically different is required to include gravity. This is where we find theories of particles without particles, brick models without bricks, called string theories.

All plastic bricks are made from the same material, their individual properties of colour and shape only resulting from a dye and a mould. The idea behind string theory suggests that the particles we observe are made, like bricks, from something more fundamental. Like the plastic of bricks, particles are made from things called strings. As plastic bricks are shaped by different moulds to form different shaped bricks, strings vibrate at different frequencies. Each different mould forms a characteristic brick. Each frequency of a string results in a different particle.

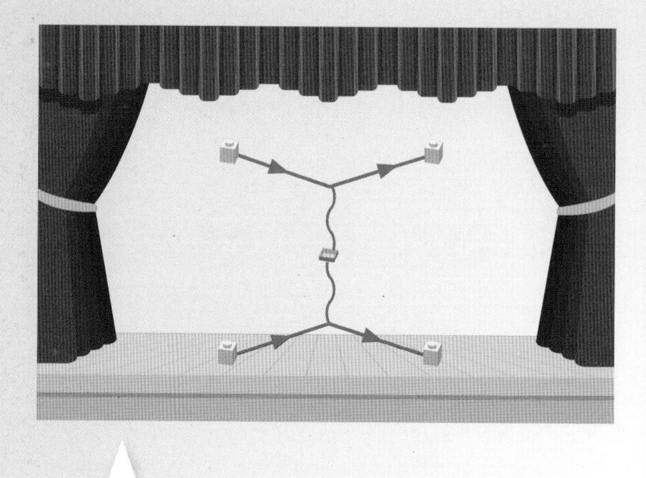

The electromagnetic, strong and weak forces of Nature determine the set routines and choreography of every particle's dance.

While we are used to living in a four-dimensional world (three dimensions of space and one of time), strings experience something altogether different. String theories exist in many more dimensions of space (some even with extra time dimensions). We cannot experience these new dimensions because they are curled so tight that they are smaller than the electrons and quarks from which we are made.

Gravity is unlike the other forces as it affects the very stage upon which particles perform by curving it.

The great divide in strength between gravity and the other forces is reconciled in string theory. The strong, weak and electromagnetic forces are exchanged in the four dimensions we are familiar with, but gravity is exchanged through all spatial dimensions. This spreads out gravity's influence more thinly so that in our four dimensions it seems weaker than expected.

MICRO BLACK HOLES

When the Large Hadron Collider was first switched on, one possible observation caused panic among many – the creation of micro black holes. Here a very rare particle interaction might tap into the extra-dimensional gravity, borrowing a huge amount of energy thanks to the uncertainty principle, to create for the briefest of moments a tear in space–time itself – a black hole. Its cataclysmic death would spew out a menagerie of particles like no other thing seen in the LHC detectors.

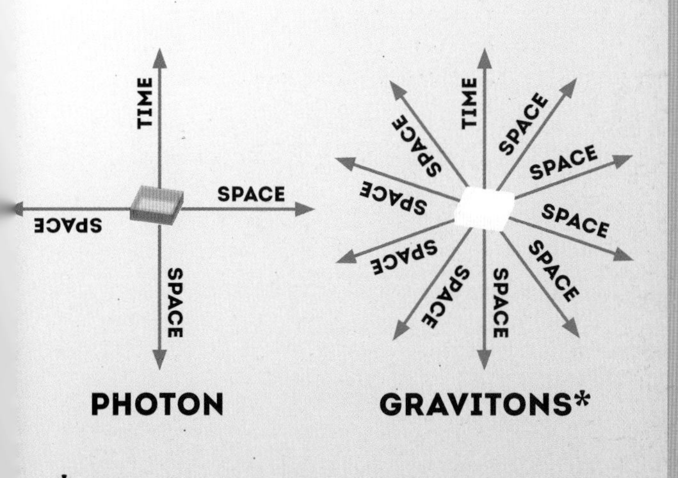

* *This particle has been hypothesised but never observed experimentally*

THE NEXT BIG STEP

Mathematicians and physicists in the early 20th century commented that an infinite number of apes typing randomly at a typewriter, given infinite time, would eventually rewrite Shakespeare's complete works. Yet with just four restrained rules imposed by forces, a limited variety of building blocks and just 13.8 billion years, similar random chance led to the production of Shakespeare himself.

Given enough plastic bricks, the rules in this book and enough time one might imagine that a plastic Universe could be built by us, brick by brick, the pinnacle of the ape world. But it would not work. The forces of Nature are set just right for life to have evolved on our small green planet. If any of the building blocks had a different size, mass or charge, then we could not exist.

Science is a progression of ever more accurate mathematical descriptions of our natural world. The strange movement of the planet Mercury could not be explained by Isaac Newton's low-energy theory of gravity – before it was complete a crack was already visible. Newton was also worried that he could explain how the planets moved but not why gravity moved them. Huge advances in other fields culminated in Albert Einstein, over 200 years later, explaining this crack with a brand new theory, General Relativity. Newton's theory explained the how but it was General Relativity that answered Newton's own questions of why gravity is felt by planets – they follow paths in a curved space–time.

In the Standard Model we have an excellent low-energy explanation of the Universe in which we live. The search continues for this theory's breaking point for a hint of what new physics lies in store. Only once cracks start appearing can we begin the future of answering the why questions – why the forces are the strength they are, the particles have the mass they do, and where our particle-filled Universe came from to begin with.

For now we can only theorize what the 'why' might be. It could be that we live in a finely tuned Universe with each different property set just so. Another idea is that we are one of an infinite number of universes each branching from one another at every particle interaction, with only ours surviving with the right conditions to create intelligent life.

Whatever the reason, we are standing on the cusp of understanding why our Universe behaves the way it does. It is time for us to move on from our Standard Model brick set and on to new things with experiments lighting our way into the dark unknown.

INDEX

ACKNOWLEDGEMENTS

This book has been in my mind since a coffee in the Senior Common Room at Queen Mary, University of London (QMUL) in 2010. I would like to thank Bryony Frost for helping turn those initial ideas into fantastic outreach resources and my other colleagues at QMUL for helping hone my plastic brick analogy over yet more coffee.

Without Trevor Davies's enthusiasm, this book would still be in my head – thank you for believing and investing yourself into the project. Thank you to Jaz and the design team for making the book look so fantastic and Sarah Green for organizing everything to run so smoothly.

I would have not survived the late nights and seven day weeks if my wife Emily had not kept me sane, thank you for your support and distracting me when I needed it. Last, and certainly not least, I would like to thank my parents, Simon and Julie, who have supported my interests and encouraged my passions from an early age – physics and everything beyond.

PICTURE CREDITS

123RF Amy Harris 94-95 background; Andrey Kryuchkov 16 left; hxdbzxy 16 right; Kari Haraldsdatter Høglund 142; nasaimages 54-55, 164; Olga Popova 19; Pavel Isupov 126-127 background; Peter Jurik 60-61 background; petkov 84-85 background; romansli 62-63; sebikus 59; ssilver 108-109; Vadim Sadovski 146-147 background; vampy1 94. **Alamy Stock Photo** Randsc 145; R Jay GaBany/Stocktrek Images, Inc. **Dreamstime.com** Graphics.vp 7, 44-45 background. **Getty Images** Bettmann 89; Graham Stuart/Stringer 138; Jamie Yan/EyeEm 171; Pascal Boegli 67; zmeel 79. **istockphoto.com** kevron2001 10-11 background, 38-39, 124-125 background. **Library of Congress** 32, 86. **NASA** ESA, and the Hubble Heritage Team (STScI/AURA) 51; JPL-Caltech/University of Wisconsin 4, 14-15 background, 24-25 background, 33 background. **Science Photo Library** CERN 114-115, 157. **Shutterstock** Andrey VP 22-23 background, 98-99 background; Aperture75 169; arleksey 30-31 background; Dmitriy Rybin 113 background, 116-117 background; GiroScience 26-27 background, 56-57.